Video Grounding and Its Generalization

Xin Wang • Xiaohan Lan • Wenwu Zhu

Video Grounding and Its Generalization

From I.D. and Task-specific Models to O.O.D. and Large Foundation Models

Springer

Xin Wang
Tsinghua University
Beijing, China

Xiaohan Lan
Tsinghua University
Beijing, China

Wenwu Zhu
Tsinghua University
Beijing, China

ISBN 978-3-031-94836-7 ISBN 978-3-031-94837-4 (eBook)
https://doi.org/10.1007/978-3-031-94837-4

This Springer imprint is published by the registered company Springer Nature Switzerland AG
The registered company address is: Gewerbestrasse 11, 6330 Cham, Switzerland

To my parents,

for unconditionally loving and supporting me. There does not exist one parenting book on earth that could have instructed you for my naughty childhood and rebellious teenage years. Thank you for always waiting for me late at night in the past years.

In fact, Dad and Mom do not read English books, so they will never know this book unless someone tells them.

Xin

Foreword by Ramesh Jain

Video has emerged as our most profound and straightforward medium for capturing and conveying experiences. Every frame pulses with life, each sequence a window into moments that would otherwise slip through time's relentless march. We store our everyday life events as well as weddings, our children's first steps, historical events, and scientific discoveries in this dynamic format—not just as passive recordings but as living memories that can transport us back to those precise moments.

For centuries, due to the limitations of technology, human knowledge has been predominantly encoded and transmitted through language. The convenience of text—its ease of creation, storage, and retrieval—has made it the default medium for preserving human thought and experience. However, as we stand at the threshold of a new era in computing, this paradigm is poised for a dramatic shift. The rise of experiential computing, where digital interactions become increasingly immersive and multi-sensory, signals a future where video will emerge as the primary vessel of human knowledge and experience.

Yet, this treasure trove of visual experience remains partially locked away. Without effective video grounding—the ability to precisely locate and understand specific elements within video content—we cannot fully harness this wealth of visual information. Like having a library without a catalog system, our video archives hold immense knowledge that remains just beyond our grasp. The challenge of video grounding isn't merely a technical puzzle; it's the key to unlocking the full potential of humanity's most powerful experiential medium, preparing us for a future where video becomes the lingua franca of human experience and knowledge sharing.

Having spent five decades in video processing, beginning in an era before digital video and powerful computing were available, I've witnessed firsthand the evolution of this field. My early conviction about the transformative role digital video would play in experience detection and sharing has been validated many times over. Thus, when I learned about this comprehensive book on video grounding, I was both excited and gratified. The timing couldn't be more perfect, as we stand at the cusp of video becoming the primary medium for human knowledge sharing.

This monograph presents a sophisticated yet accessible framework for addressing one of the most challenging aspects of video understanding: the leap from controlled, in-distribution scenarios to the complexity of real-world applications. The authors' unified treatment expertly bridges theoretical foundations with practical implementations, making it an invaluable resource for both researchers and practitioners.

Wenwu Zhu, a pioneer in multimedia computing and big data, and Xin Wang, an emerging leader in machine learning and video analysis, bring together their rich expertise to craft this comprehensive work. What makes this book stand out is its balanced approach: while deeply technical and rigorous, it remains accessible and practical. The authors interweave theoretical insights with real-world applications, offering readers a pathway to understand both the "why" and the "how" of video grounding. From academic researchers to industry practitioners, this monograph serves as a valuable resource for those seeking to develop next-generation video understanding systems.

As we move toward a future where video becomes the primary medium for capturing and sharing human experience, the principles and techniques presented in this book will become increasingly crucial. This work not only advances our technical understanding but also helps pave the way for more intuitive and natural ways of interacting with our growing video archives.

This book is more than a technical guide; it is an inspiration for the future of video understanding. I am delighted to present this seminal work to both current and future generations of researchers and practitioners—a book that I believe will play a pivotal role in transforming how we understand and harness the power of video in shaping human knowledge and experience.

Irvine, CA, USA
January 2025

Ramesh Jain
Emeritus Bren Professor
Department of Computer Science
University of California, Irvine

Foreword by Tat-seng Chua

In recent years, video understanding has become one of the most critical and challenging tasks in artificial intelligence, with wide-ranging applications in areas such as surveillance, content recommendation, autonomous vehicles, and human-computer interaction. The task of video grounding, which involves aligning video content with specific queries, actions, or concepts, has emerged as a key problem in this domain. It is a problem that requires not only deep expertise in machine learning but also a nuanced understanding of how multimedia data—especially video—can be effectively analyzed and understood.

This book, *Video Grounding and Its Generalization*, authored by Xin Wang and Wenwu Zhu from Tsinghua University, provides a timely and comprehensive exploration of video grounding and its evolution, from in-distribution (ID) to out-of-distribution (OOD) settings. Furthermore, as the field moves from domain-specific models to more versatile, foundation models, the ability to adapt these models to a wide array of video tasks becomes increasingly important. The authors address this challenge by examining how current models perform in controlled, in-distribution environments and how they can be adapted to work in more dynamic, real-world situations where video data is often unpredictable and noisy.

Xin Wang and Wenwu Zhu are both highly respected scholars whose expertise in machine learning, multimedia computing, and video analysis makes them ideally suited to tackle this topic. They have successfully combined their deep technical knowledge with practical insights, providing a rich account of the state-of-the-art in video grounding. This book is structured to offer both foundational knowledge and practical methodologies, covering a range of techniques that can handle video data's inherent complexity and variability.

This book is not only a comprehensive reference for those working in video understanding but also an important contribution to the broader field of AI and machine learning. It will be valuable to researchers, practitioners, and graduate students interested in the application of AI to multimedia data, and those who are keen on understanding how AI models can be generalized beyond their initial training conditions.

The work of Xin Wang and Wenwu Zhu represents a significant step forward in our understanding of video grounding, and I highly recommend this book to anyone seeking a deep, technical understanding of this rapidly advancing area of artificial intelligence for videos.

Singapore, Singapore
January 2025

Tat-seng Chua
KITHCT Chair Professor
School of Computing
National University of Singapore

Foreword by Jiebo Luo

Machine learning (ML) has transformed our understanding of and interaction with the world, driving innovations across a wide array of fields such as computer vision and natural language processing. In particular, video grounding—the task of associating video content with textual queries—stands as one of the most fascinating and complex challenges in the realm of multimedia intelligence. As video data becomes ubiquitous and integral to many applications, the ability to effectively understand and interact with videos through automated systems is critical for revolutionizing application domains ranging from autonomous vehicles, smart surveillance, to virtual assistants, and beyond.

The journey toward solving video grounding is marked by a transition from simpler, task-specific models to the more generalizable, large-scale models that dominate today's AI landscape. Early models, designed for specific domains or well-defined datasets, often struggled with generalization. This paradigm, however, has started to shift with the advent of foundation models—large, pre-trained models that can adapt across a broad range of tasks. The shift from in-distribution (ID) tasks to out-of-distribution (OOD) problems, however, is a formidable challenge that requires nuanced understanding, effective model design, and novel strategies for training and evaluation.

In this monograph, Xin Wang and Wenwu Zhu, two leading scholars from Tsinghua University, embark on an in-depth exploration of this transition, providing a comprehensive overview of the development of video grounding technologies and their subsequent generalization to more dynamic and diverse tasks. Their work focuses on a critical path: moving from models that excel within narrow, controlled environments to those capable of addressing the complexities of real-world, out-of-distribution scenarios. This focus is not only technical but also philosophical, reflecting the deeper evolution of AI itself—from systems designed for highly specific tasks to systems that must generalize, adapt, and interact with the world in more human-like ways.

Xin Wang's expertise in machine learning and multimedia analysis, combined with Wenwu Zhu's deep experience in multimedia computing and networking, creates a powerful synergy that is evident throughout the book. The authors not only explore the cutting-edge techniques and methodologies behind video grounding but also provide insights into the broader implications of these advancements for AI and its applications.

This monograph is timely for anyone interested in the future of video understanding, AI generalization, and the broader evolution of machine intelligence. It offers an accessible yet deep dive into the current state of the art, challenges that lie ahead, and the exciting opportunities that emerge as video grounding evolves. For researchers, practitioners, and students, this work will serve as both a guide and an inspiration, sparking new ideas and facilitating the growth of this rapidly advancing field.

I am confident that this book will contribute significantly to the ongoing discourse in machine learning, serving as a foundational reference for current and future research in video grounding and its generalization. Xin Wang and Wenwu Zhu have masterfully captured the essence of their research journey in this research area. This book is highly recommended to anyone interested in understanding the cutting-edge development in vision and language.

Rochester, NY, USA
January 2025

Jiebo Luo
Albert Hopeman Professor
Department of Computer Science
University of Rochester

Preface

We are more than happy to write this monograph for video grounding, where the initial motivations stem from the following considerations.

First and foremost, the field of video grounding is rapidly evolving, particularly in light of the recent advancements in large foundation models and their ability to analyze videos. While traditional models have achieved remarkable success in specific tasks with well-curated datasets, the transition to more generalizable and even scalable solutions calls for a comprehensive monograph that bridges this critical knowledge gap via providing an overall summary and deep insights for understanding and extending these achievements. This book offers a structured narrative on the journey from in-distribution and small task-specific models to out-of-distribution generalization and large foundation models within video grounding.

Second, the intersection of video grounding and large foundation models represents a paradigm shift in artificial intelligence research and applications. Existing perspectives often focus either on the technical intricacies of task-specific models or on the broader potential of foundation models without bridging the two. This book seeks to unify these viewpoints by examining how foundational models can generalize across diverse tasks while maintaining their effectiveness in specific scenarios.

Third, we dedicate significant attention to the practical implications and recent breakthroughs in this domain. Topics such as multimodal fusion, disentangled representation learning, curriculum learning, and the utilization of foundation models in video grounding are discussed in detail, reflecting cutting-edge progress in addressing problems such as debiasing and versatility, etc. By incorporating discussions on these recent advances, we aim to provide a forward-looking perspective that aligns with the broader trajectory of machine learning research.

Furthermore, we emphasize the practical applications of video grounding in a wide range of fields, including computer vision, robotics, and multimedia analysis. By highlighting both theoretical foundations and real-world implementations, we aspire to make this book valuable to an interdisciplinary audience, from academic researchers to industry practitioners. Last but not least, we are committed to addressing the needs of a diverse audience, designing this book to serve as both an introductory guide for students and early-stage researchers and a valuable reference for experienced experts in academia and industry. It balances foundational knowledge with state-of-the-art techniques, making it accessible yet comprehensive.

We sincerely hope that readers may find this book an insightful and thought-provoking resource, whether they are embarking on their first steps in video grounding research or are seasoned professionals looking to deepen their understanding of this dynamic field.

Beijing, China Xin Wang
Beijing, China Xiaohan Lan
Beijing, China Wenwu Zhu
January 2025

Acknowledgments

We thank Yitian Yuan for initiating our research on video grounding during her graduate study. Many thanks to Bin Huang, Benyuan Meng, Wei Feng, and Houlun Chen from Tsinghua University for their contributions during the writing of this book.

We would also like to express our sincere thanks to Ramesh Jain from University of California, Irvine, Tat-seng Chua from National University of Singapore, and Jiebo Luo from University of Rochester for writing forewords for this book. This monograph will never be finished without their valuable inputs.

This monograph is supported by the National Natural Science Foundation of China No. 62222209, Beijing National Research Center for Information Science and Technology under Grant No. BNR2023RC01003, BNR2023TD03006. Wenwu Zhu is the corresponding author.

Competing Interests The authors have no competing interests to declare that are relevant to the content of this manuscript.

Contents

List of Figures

List of Tables

Part I
Methodologies for Video Grounding

Part I will cover basic and advanced methodologies for *Video Grounding*, discussing key comparisons with several representative Vision-Language learning tasks including multimodal understanding and generation.

Chapter 1
Introduction

Abstract In this chapter, we begin by introducing the basic concepts and technical advancements for *Video Grounding*. We highlight its importance within the research community, discuss its relationships with other Vision-Language learning tasks, and share our insights on the potential extensions towards more generalized settings.

1.1 Video Grounding

Video Grounding, as one important Vision-Language learning task also known as temporal sentence localization in videos, which aims to localize one target segment from an untrimmed video regarding a given sentence query, has drawn increasing attention in the research community over the past few years. As shown in Fig. 1.1, given the query "A little girl walks by a little boy and continues blowing the leaves" as input, the goal of *Video Grounding* is to predict the start and end points (i.e., 7.11–12.7 s) of the target segment within the whole video, and the predicted segment should contain the activities indicated by the input query.

This task has become especially important with the rapid development of mobile devices and multimedia technologies, which have made video content more accessible. As visual and audio content convey more complex information compared to texts and images, videos contain richer semantics. In addition, natural language captions substantially complement video content. However, raw videos on the Internet are usually redundant and yet fail to satisfy user-specific retrieval demands, consuming a large amount of storage resources. Therefore, it is vital to localize segments based on users' interests, which also benefits storage optimization and promises to make *Video Grounding* more popular in the academic community.

Similar to other visual-language learning tasks (e.g., visual question answering [6, 7, 235], image/video captioning [13, 169, 170, 217, 231, 232], visual grounding [93, 207, 234] and vision-and-language pre-training [15, 30, 148, 197]), *Video Grounding* requires understanding of both visual and textual input. Moreover, it can serve as an intermediate task for various downstream vision-and-language tasks such as video question answering [43, 110, 115, 218] and video summary [28, 156, 183, 238, 254]. For example, related segments can be first grounded through

X. Wang et al., *Video Grounding and Its Generalization*,
https://doi.org/10.1007/978-3-031-94837-4_1

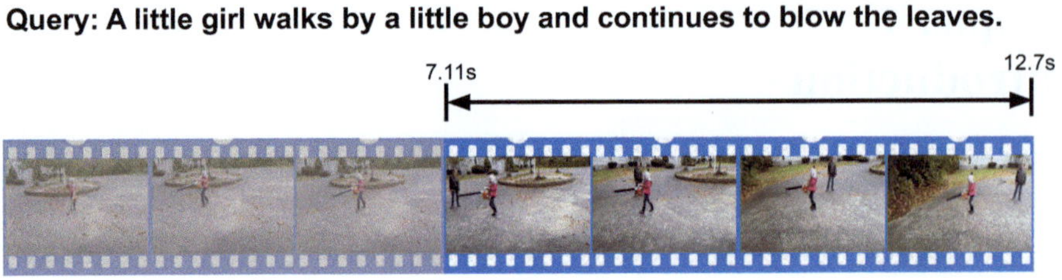

Fig. 1.1 *Video Grounding* aims to predict the start and end points of the target segment within the whole video with respect to a given sentence query

the textual question and then analyzed for discovering the final answer to the input question. Also, by providing concise textual summaries of videos, semantically coherent video segments can be grounded, retrieved, and composed as visual summaries of the original videos.

As such, we argue that *Video Grounding* is an emerging and challenging task in the fields of computer vision and natural language processing. It involves locating a specific temporal segment that corresponds to a given natural language query within a video. This task bridges the gap between visual content and linguistic descriptions, enabling machines to understand and interpret videos in a more human-like manner. The idea is to align textual descriptions with precise temporal intervals in a video, effectively grounding language in visual data. The motivation behind *Video Grounding* stems from the exponential growth of video content on the internet and the need for efficient methods to analyze, search, and retrieve relevant information from these vast datasets. Traditional video analysis techniques often rely on metadata or manual annotations, which are time-consuming and impractical for large-scale applications. *Video Grounding* automates this process by leveraging the semantic relationship between visual and textual modalities, allowing more intuitive and accurate video understanding. This capability is particularly important in applications such as video summary, content-based video retrieval, human-computer interaction and assistive technologies for the visually impaired. The importance of *Video Grounding*, lies in its potential to enhance the way machines interact with and interpret multimedia content. By enabling precise localization of events described in natural language, it facilitates more sophisticated video analysis and reasoning. For instance, in surveillance systems, video grounding can help identify specific incidents described in textual reports. In educational platforms, it can assist students in locating relevant segments of instructional videos based on their queries. Moreover, it plays a crucial role in advancing multi-modal AI systems, where understanding the interplay between vision and language is essential for tasks like video captioning, question-answering, and storytelling.

To summarize, *Video Grounding* represents a significant step toward achieving fine-grained video understanding by linking textual descriptions to specific temporal segments in videos. Its development is driven by the need for efficient video analysis tools and the growing demand for intelligent systems capable of interpreting multi-modal data. As research in this area progresses, *Video Grounding* is expected to unlock new possibilities for applications across diverse domains, making it a vital area of study in AI and multimedia research.

In this chapter, we will first introduce the preliminaries about Vision-Language learning and discuss the relations between video grounding and other Vision-Language tasks. Then we briefly talk about the possible extensions towards Generalized *Video Grounding*.

1.2 Other Representative Vision-Language Learning Tasks

Humans perceive the physical world primarily by visual information, then they organize and express thoughts via language in brains. This process serves as the foundation of human intelligence and inspiration to artificial intelligence (AI).

Computer vision (CV) and natural language processing (NLP) have been classic research areas that aim to simulate the above two characteristics respectively [119], and have made significant progress due to the boost made by deep learning in recent years. In spite of the advances in uni-modality domains, tackling real-world problems still requires integrating multiple modalities, especially vision and language, to achieve human-like comprehension.

The integration of vision and language in multi-modality research is gaining attention for its role in visual perception, which is essential for understanding the environment. At the same time, learning good language-aligned visual features greatly improves the performance of both vision and vision-language tasks. Moreover, the availability of extensive datasets and benchmarks has contributed to the popularity of vision-language learning as well.

In general, most approaches to Vision-Language learning mainly focus on task-specific modeling, where models are well-tailored for specific problems. Though these methods show good performance in target tasks, it is difficult to transfer from one model to others. To make the whole book self-contained, we next discuss task-specific Vision-Language learning approaches via categorizing them into image-based methods and video-based methods.

1.2.1 Image-Text Understanding

For image-text understanding, it learns correlation between the image and text modalities to complete downstream tasks, which contains the following typical tasks, i.e., Image-Text Matching (ITM), image Visual Question Answering (iVQA), Image Captioning (IC), and Visual Grounding.

Image-Text Matching (ITM) ITM aims to match the semantically closest instance from one modality given a query of another modality, where modalities are images

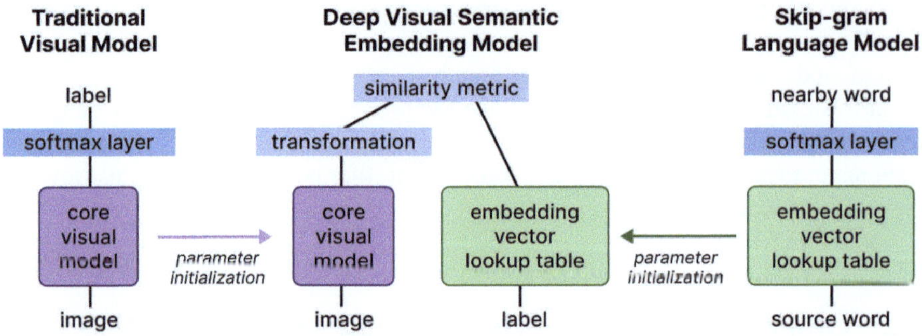

Fig. 1.2 The framework of DeViSE [51] where a deep visual semantic embedding model uses partially initialized parameters from a skip-gram language model and a traditional visual model to bridge the gap

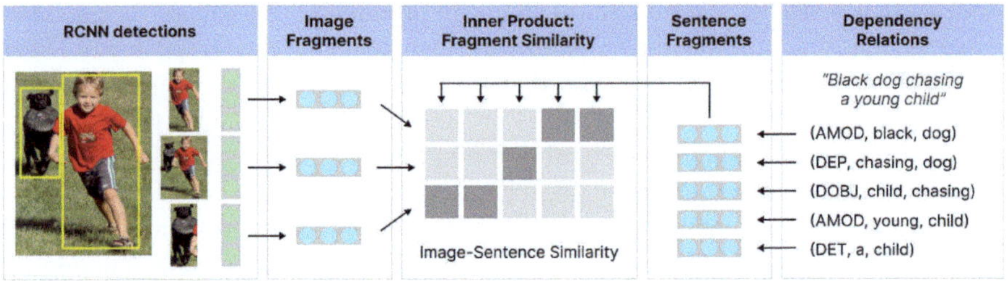

Fig. 1.3 The architecture of Deep Fragment [92]. **Left**: CNN representations of detected objects are mapped to the fragment embedding space. **Right**: Dependency tree relations in the sentence are embedded

or texts, generally including image-to-text retrieval and text-to-image retrieval tasks. ITM requires quantifying the similarity of an image and a paragraph of text.

Early ITM works mainly use global semantic representations for image and text modalities. However, DeViSE [51] (c.f. Fig. 1.2) transfers semantic knowledge learned in the text domain to the visual representation. This can be attributed to its connection between a simple neural language model and a state-of-the-art visual task-oriented model and parameters inheriting from both to improve visual object recognition, which can be seen as an image-to-text retrieval task. DT-RNN [190] introduces dependency trees to embed sentences into a vector space for better image retrieval. VSE++ [42] employs hard negative mining to generate better visual-semantic embeddings.

As it's not enough to cover important details of multiple modalities merely from global representations, Karpathy et al. [92] propose Deep Fragment (c.f. Fig. 1.3), which is the first to use fine-grained representation for images and texts. Specifically, they use a similarity matrix to align fragments between various modalities in the cross-modality embedding space, modeling dependency tree relations and object detection as sentence and image fragments, respectively.

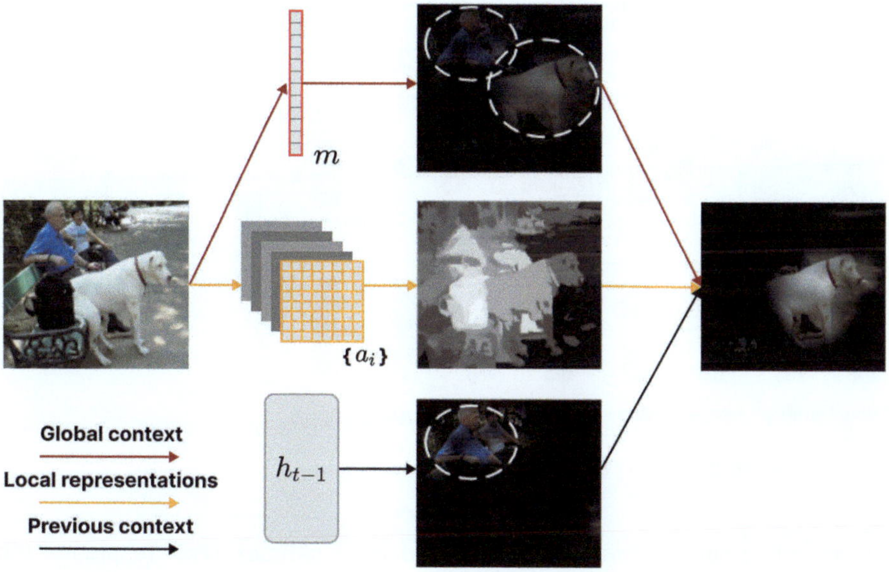

Fig. 1.4 Illustration of the context-modulated attention mechanism proposed by Huang et al. [75]

As the attention mechanism has prevailed in vision or language tasks, Huang et al. [75] introduce attention into ITM research. They design a context-modulated attention scheme (c.f. Fig. 1.4) in a selective multi-modal LSTM to attend to image-text instance pairs in a few timesteps, where multiple cross-modal local similarities are aggregated to obtain a final matching score. Nam et al. [165] use visual and textual attention mechanisms to model fine-grained interplay, where they focus on specific regions in multiple steps to gather important information. Lee et al. [108] present Stacked Cross Attention to discover the full latent alignments using a set of images and sentences as context, both of which are used as queries attending to each other.

The core of task-specific ITM methods is to map the images and texts to a shared semantic space where the matched pair of images and texts should be well aligned. With fine-grained interaction of component-level information in both modalities, better semantic alignment can be achieved.

Image Visual Question Answering (iVQA) Given an image-question pair, iVQA aims to answer this question based on the given image. iVQA methods can be classified depending on the output formats, i.e., whether it is a fixed set of candidates or free-formed answers.

Vanilla iVQA, first proposed by Antol et al. [7] (c.f. Fig. 1.5), formulates the task as classification in multiple candidate choices and combines an LSTM-based [70] question encoder and a VGG-based [188] image encoder. Image embedding and question embedding are fused by point-wise multiplication. Then, the fused features

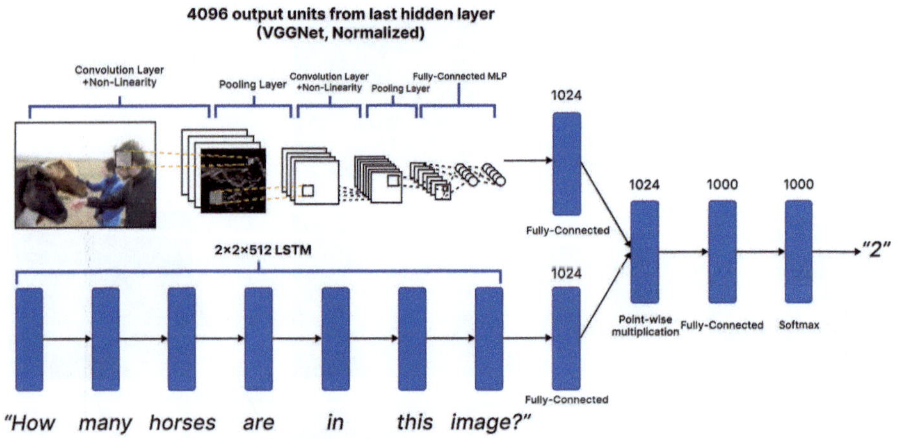

Fig. 1.5 Pipeline of vanilla iVQA [7]

are passed through a linear layer followed by a softmax layer to predict the probability of candidate answers.

In addition to cross-modal comprehension, iVQA calls for image reasoning in response to the provided query. Subsequent iVQA works mainly follow the same prototype with LSTM for sequential modeling. Some studies fuse image information during the encoding or decoding process to execute better reasoning on the image. Malinowski et al. [157] propose to feed the CNN image feature into each LSTM cell of the question encoder. Gao et al. [57] fuse the CNN image feature with the output of each decoder cell and generate the answer word by word.

However, early works [7, 57, 157] mainly adopt global representation for images as well as a simple fusion strategy, which fails to attend to crucial clues in images for answer prediction. The reason is that redundancy exists in images which may lead to noisy information.

Similar to ITM, the attention mechanism has been widely used in iVQA. SAN [225] is the first work to verify the effectiveness of attention in iVQA via proposing to stack multiple question-guided attention layers (c.f. Fig. 1.6) to reason progressively. In each layer, the question's semantic representation is used as a query to search for related image grids. Fukui et al. [54] also use image grid features and improve the fusion strategy of image and question features through multi-modal compact bilinear pooling [130].

Grid features have limited power because grids align poorly with objects, which closely correlate to question semantics. Thus, many works capture local representations of images in an adaptive fashion, by being attended by the question. Shih et al. [187] design a region selection layer (c.f. Fig. 1.7) to selectively identify informative image regions from candidates proposed by Edge Boxes [261]. BUTD [6] adopts a pre-trained Faster-RCNN [179] detector to generate candidate object-level region features and uses the question features as query to attend. Based on the assumption that attention over text is of equal importance as attention over

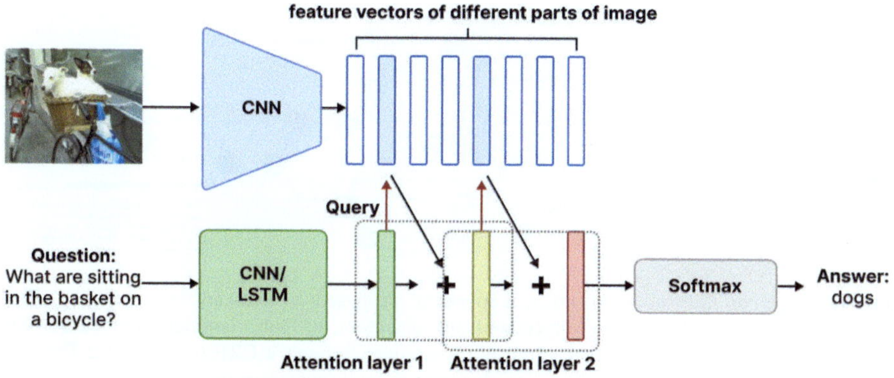

Fig. 1.6 Stacked Attention Network (SAN) for iVQA by Yang et al. [225]

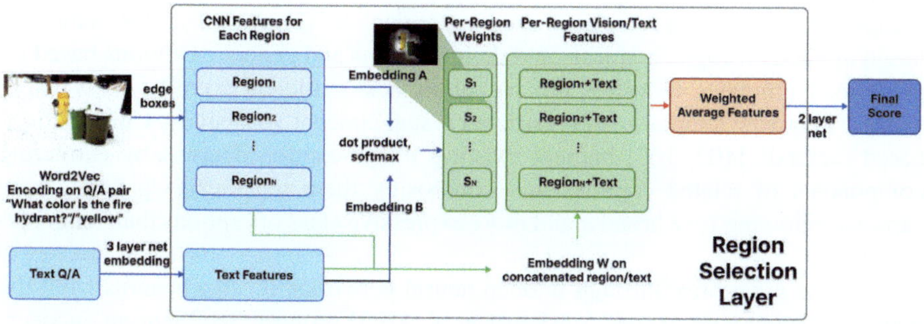

Fig. 1.7 Details in region selection layer by Shih et al. [187]

the image, Lu et al. [145] develop a co-attention method to perform mutual attention jointly between texts and images.

Although there are other modality fusion strategies [10, 95, 178], attention is unquestionably the most popular fusion strategy.

iVQA can be seen as a "soft grounding" task on the image where the models are required to find areas related to the question explicitly or implicitly, involving recognizing objects, extracting relations, and calculating relevance. The core of iVQA is to obtain a joint representation of image and language. We argue that the goal of iVQA is also applicable to *Video Grounding* and techniques in iVQA can somehow be transferred to *Video Grounding*.

Image Captioning (IC) The goal of image captioning is to generate a caption for a given image. The caption is expected to summarize the main content of the image, which usually contains salient objects/humans, activities, and positional relations. Different from iVQA, the IC task requires flexible and comprehensive output that reflects the image information as much as possible, other than focusing on local information.

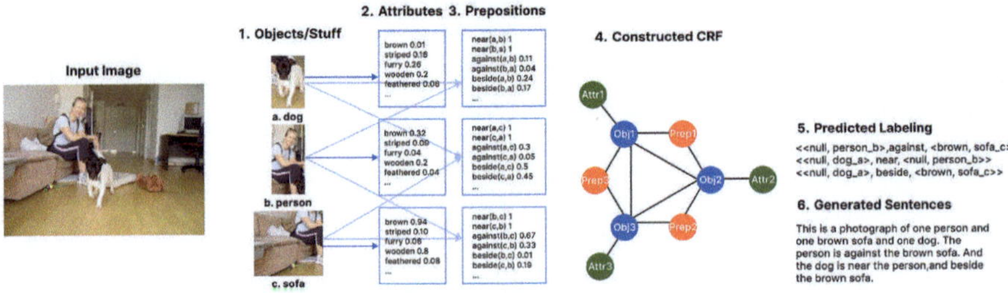

Fig. 1.8 Pipeline illustration of image captioning proposed by Kulkarni et al. [101]. They propose a framework that includes object and stuff detectors, attribute classifiers, and relationship prediction functions. A conceptual graph is constructed through a CRF (Conditioned Random Field) and the corresponding labeling is predicted to generate sentences by rules

The majority of the early IC techniques [46, 101] are rule-based, including composition-based and template-based strategies. As shown in Fig. 1.8, template-based methods recognize objects and their relations and generate captions based on predefined templates. However, this paradigm lacks sufficient expressiveness due to its rigid language templates and restricted visual concept vocabulary. Composition-based methods [102, 103] harness existing image-caption datasets by extracting components of related captions and composing them together to generate new captions, allowing for a broader and more expressive class of captions than template-based ones.

With the great breakthrough in deep neural networks, IC also benefits from the power of deep architectures. Sutskever et al. [195] propose Seq2Seq, an encoder-decoder structure, which achieves great success in machine translation. Formulating the IC task as an image-to-text translation, Kiros et al. [99] (c.f. Fig. 1.9) design an encoder-decoder pipeline for image captioning which learns cross-view representation with a hinge-based triplet ranking loss. Xu et al. [216] (c.f. Fig. 1.10) replace the text encoder in the translation task with an image encoder GoogleNet [196]. Following them, many later works use this img2seq framework, where the image is encoded usually by a CNN and the caption is generated by a decoder such as the widely used LSTM. From the perspective of network connections, the CNN features can be supplied to the LSTM decoder as the initial hidden state [91, 205, 216] or to each LSTM cell in the decoding process [37, 159].

A similar problem in IC is that global CNN features fail to focus on more informative regions. To address this issue, attention is brought, Xu et al. [216] propose to enhance the decoding process with attention in each LSTM cell, where the hidden state attends to grid features to select important ones in the current temporal step. Since splitting the image into equally-sized grids ignores semantics in it, later works improve it through flexible region selections. BUPT [6] aligns attention with salient regions acquired with a detection model, which benefits both IC and iVQA.

There are several variations regarding how to attend. For example, Lu et al. [146] claim that the decoder does not need to always attend to visual features since

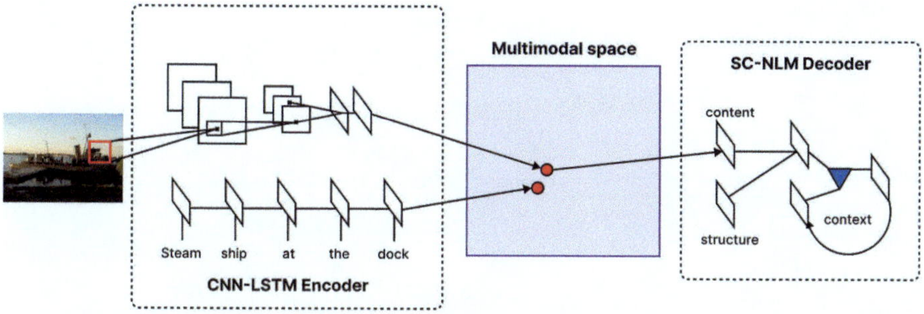

Fig. 1.9 The architecture proposed by Kiros et al. [99]. Encoder: A CNN and LSTM for learning a joint image-sentence embedding. Decoder: A new neural language model that combines structure and content vectors for generating one word at a time in sequence

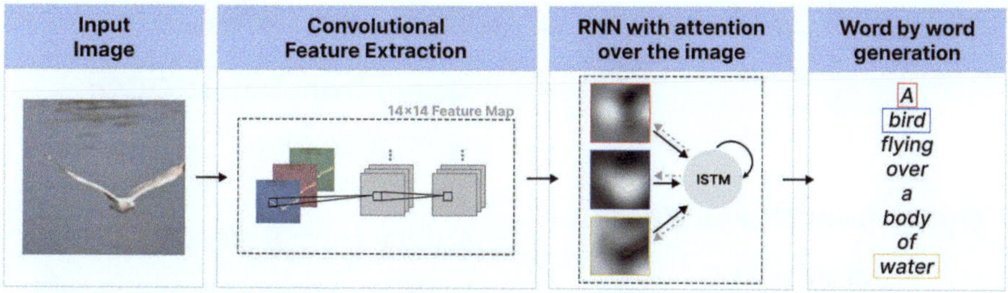

Fig. 1.10 The pipeline proposed by Xu et al. [216]. It learns a word-image alignment by LSTM with attention to the image in the word-by-word generation

some words are irrelevant to visual features. As a result, they suggest using a gate to determine attendance. Huang et al. [76] design a special mechanism called "Attention on attention" (AoA) for the image captioning task. After standard attention, AoA concatenates the attended vector and the query. Then, it generates an information vector and an attention gate from the concatenated vector and multiplies the gate and information vector to obtain the output.

Except for works with attention mechanisms, there are also works that do not use attention. For example, Lu et al. [147] first generate a sentence template and then fill it with concepts detected in the image. Cornia et al. [147] generate a sentence by predicting a sequence of noun chunks. They first detect regions and then sort the regions with a sorting network. Finally, each region will be converted to a noun chunk to form the sentence.

In summary, early image captioning methods develop along two main aspects: visual representation and language decoding. Visual representation evolves from image-level global features to fine-grained and object-level region features, and language decoding develops from LSTM to attention-based models.

Visual Grounding aims at locating object regions corresponding to the noun descriptions in an image. Unlike object detection, which only aims to detect the

Fig. 1.11 Visual grounding. The target of the input query is an instance in the image, and the input query itself may be a dialogue, a sentence, or a phrase

objects or the regions of interest given the pre-defined class labels, visual grounding expects to understand the natural language query and then find out the target object of the query in the image, which can be widely used in the visual understanding system and dialogue system (Fig. 1.11).

The initial visual grounding methods apply two-stage approaches, which learn the bounding box of the referent object by the cross-modal matching between a set of region proposals (e.g., Fast-RCNN) and the language features. Hu et al. [73] propose three LSTMs to process linguistic, local, and global information, respectively, considering visual grounding as a generalization of object detection. Deng et al. [32] propose an accumulated attention mechanism to reason among the textual query and grounding image jointly, circularly accumulating the attention for useful information in the image, query and objects while gradually ignoring noises.

Meanwhile, anchor-based one-stage methods predict the bounding box according to the maximal confidence over dense anchors and the fused two-modality features. Considering that generating appropriate proposals in the zero-shot grounding setting would be an obstacle as the proposal generator is trained on the entities common in the detection and grounding datasets, Sadhu et al. [181] propose a single-stage model called Zero-Shot Grounding Network that combines the detector network and the grounding system which predicts classification scores and regression parameters. However, these detection-based one-stage approaches still extract localization on-grid features individually, lacking the contextual information in the whole scene, such as the relationships between objects. To address these issues, Huang et al. [78] place emphasis on the relation modeling in one-stage visual grounding and propose a novel LBYL-Net to model long-range and spatial-aware features with Dynamic Max Pooling and Landmark Feature Convolution. In addition, to liberate one-stage grounding methods from the pre-defined dense anchors, many researchers propose anchor-free one-stage methods that usually rely on transformers that directly regress the bounding box coordinates based on the given expression. Deng et al. [33] first propose TransVG for the end-to-end visual grounding to avoid overfitting to datasets with specific scenarios. To reduce the heavy dependence on human annotations, Jiang et al. [87] propose a novel Pseudo-Q to generate pseudo-language queries for supervised training automatically. Despite their encouraging visual grounding results, these approaches frequently ignore the context's discriminative information. Moreover, they are primarily designed for standard scenes and struggle to generalize to open-vocabulary scenes. Inspired by the large-scale Vision-Language Pre-training (VLP) which shows promising performance in the open-vocabulary scene, Tang et al. [198] propose a robust visual grounding framework TransCP, shown in Fig. 1.12, which consists of four modules including a two-branch encoder, a context disentangling module, a prototype discovering and inheriting module, and a box regression module, being able to handle both standard and open vocabulary visual grounding tasks.

Visual grounding datasets include ReferIt, Flickr30k Entities, RefCOCO/RefCO CO+/RefCOCOg and Ref-Reasoning. ReferIt [93] contains 20,000 labeled images collected from the SAIAPR-12 [41] dataset which includes some uncertain queries such as "any", "whole", and some wrongly labeled data. ReferIt is divided into three subsets for training, validation, and testing the visual grounding task. These subsets comprise 54,127, 5842, and 60,103 queries, respectively. Flickr30k Entities [173] augments 158k captions from Flickr30k [64] with 244k co-reference chains linking mentions of the same entities in images as well as 276k manually annotated bounding boxes corresponding to each entity, making it essential for continued progress in automatic image description and grounded language understanding. Each image in Flickr30k Entities is attached with five refer expressions. 31,783 images and 427k short referred phrases rather than long sentences are contained in this dataset. RefCOCO, RefCOCO+, RefCOCOg all come from the MSCOCO dataset [129]. RefCOCO contains 50,000 objects from 19,994 images with 142,209

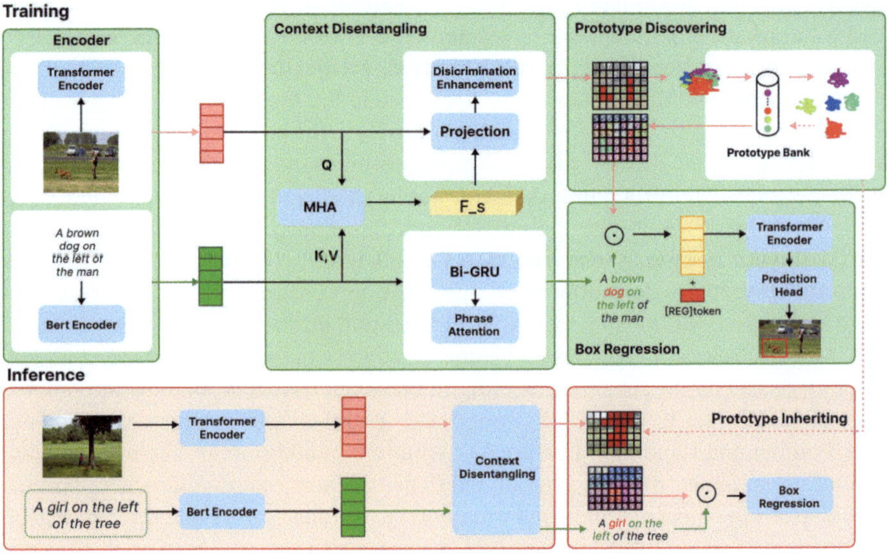

Fig. 1.12 Illustration of robust visual grounding based on Transformers (TransCP)

referring expressions and is split into training, validation, testA, and testB subsets with 120,624, 10,834, 5657, and 5095 expressions, respectively. The expressions of testA mostly refer to humans while the expressions of testB are mainly related to various objects except humans. RefCOCO+ contains 141,564 referring expressions attached to 49,856 objects from 19,992 images. Different from RefCOCO, the RefCOCO+ dataset does not take the absolute locating expressions such as "right", "top", and "besides" into consideration. Instead, RefCOCO+ pays more attention to the appearance attributes of referents. RefCOCO+ is also split into training, validation, testA, and testB with 120,191, 10,758, 5726, and 4889 expressions respectively. RefCOCOg has 25,799 images with 49,822 objects and 95,010 corresponding referring expressions. This dataset includes two split versions e.g., RefCOCOg-google [160] and RefCOCOg-umd [164], and RefCOCOg-google is for the open-vocabulary evaluation. Ref-Reasoning [226] is a large-scale real-world dataset built on GQA [83] for structured referring expression reasoning, which includes 791,956 referring expressions in 83,989 images. These referring expressions are automatically generated based on scene graphs from the visual Genome [100] dataset by a set of expression templates. The Ref-Reasoning dataset contains 1664 object classes, 308 relation classes, and 610 attribute classes. Meanwhile, the generated expressions are extremely complex with each expression describing a maximum of five objects and their relationships. There are 721,164, 36,183, and 34,609 image-expression pairs in the training, validation, and test sets of the dataset, respectively.

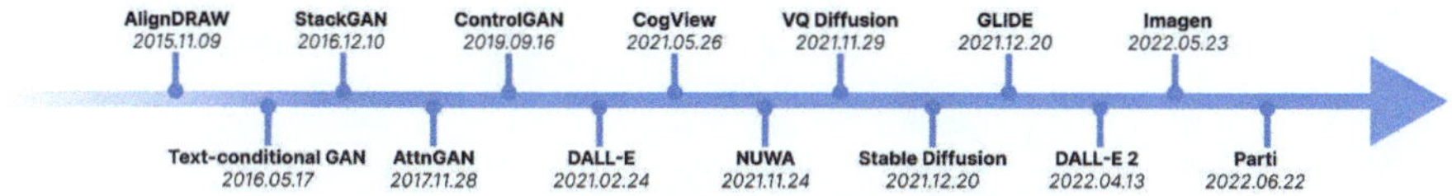

Fig. 1.13 Representative works on text-to-image task over time [248]

1.2.2 Text-to-Image Generation

Understanding and generation tasks are closely related in multi-modal learning [23]. T2IG is the task of generating images from text, which can be seen as the reverse operation of image captioning [40]. T2IG is an important task of conditional image generation, where methods take more inputs except for the images and regard them as the condition to specify the image target. More difficult than text generation, conditioned image generation is required to balance both the image's conformity to the condition and the aesthetics.

Figure 1.13 [248] shows representative methods in T2IG. Although Aligh-DRAW [158] is a groundbreaking work, its ability to produce images from natural languages is constrained by impractical outcomes.

After that, GAN [65] or conditional GAN (cGAN) [162] architectures rise up swiftly. Generative Adversarial Networks (GANs) [65] are a class of deep learning models consisting of two neural networks: a Generator and a Discriminator. While the Discriminator seeks to discern between actual data and data produced by the Generator, the Generator seeks to fool the Discriminator by producing realistic data. This adversarial training pushes the Generator to produce more realistic data. The loss function of a GAN is defined by the loss of both the Generator and the Discriminator:

$$\min_{G} \max_{D} V(D, G) = \mathbb{E}_{x \sim p_{\text{data}}(x)}[\log D(x)] + \mathbb{E}_{z \sim p_z(z)}[\log(1 - D(G(z)))], \quad (1.1)$$

where:

- G is the Generator, which tries to generate realistic data $G(z)$.
- D is the Discriminator, which tries to distinguish between real data x and generated data $G(z)$.
- $p_{\text{data}}(x)$ is the distribution of the real data.
- $p_z(z)$ is the input noise distribution for the Generator.

The training process of a GAN involves this adversarial optimization, leading the Generator to eventually produce highly realistic data that is hard for the Discriminator to differentiate from real data.

Mirza et al. [162] propose cGAN by adding condition inputs to the generator and discriminator to realize controllable generation. Sharma et al. [184] propose an approach that combines VisDial dialogues and text descriptions to generate images, inspired by StackGAN. Xu et al. [219] design the Attentional Generative

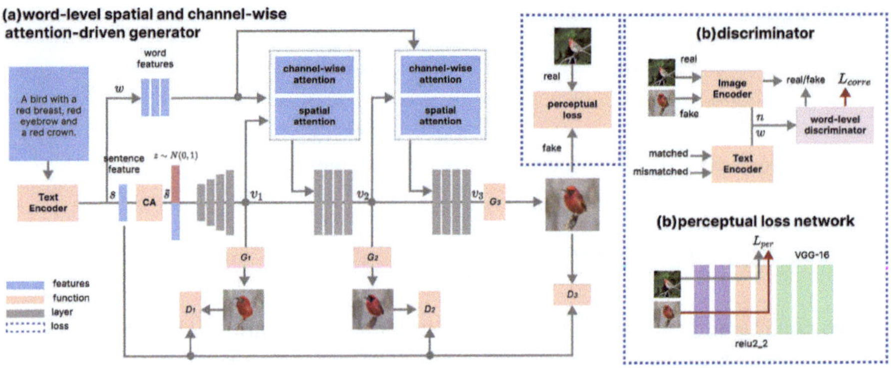

Fig. 1.14 ControlGAN's architecture. It has a word-level spatial and channel-wise attention-driven generator and a discriminator equipped with a correlation loss and a perceptual loss

Adversarial Network (AttnGAN), which synthesizes detailed image sub-regions by attending to relevant words in the natural language description. Another representative work is ControlGAN [114] (c.f. Fig. 1.14), which provides control over image generation at individual visual attributes for high-quality results. Zhang et al. [246] propose Cross-Modal Contrastive Generative Adversarial Network (XMC-GAN) and maximize the mutual information between image and text to generate more coherent, photo-realistic images with higher semantic fidelity to the conditioned text. GigaGAN [90] scales up GANs to benefit from larger datasets. Please refer to more works [2, 79, 171, 174, 199, 233, 242, 251] that propose solutions based on GAN.

However, a common problem with all GAN-based techniques is that GAN is unstable and challenging to train. Besides, they are mainly in the small-scale data regime [248]. Different from GAN-based methods, autoregressive methods [36, 176, 214, 237] treat T2IG as a seq2seq problem and exploit large-scale data for T2IG, such as OpenAI's DALL-E [176] and Google's Parti [237]. DALL-E [176] autoregressively models the text and image tokens as a single type of data by a transformer and reconstructs images from the discrete VAE. DALL-E does well in zero-shot cases. Parti [237] combines a transformer-based encoder-decoder with ViT-VQGAN, which encodes images as sequences of discrete tokens, as shown in Fig. 1.15. Parti performs better when models are scaled up, and large models do better in scenarios when external knowledge is needed.

More recently, the diffusion model (DM) has become increasingly popular in image generation. It introduces Gaussian noise continuously to disrupt the training data and then recovers the data through learning noise prediction. During inference (sampling), the diffusion model uses the learned denoising process to generate data from randomly sampled noise. Formally, diffusion models [191] are a class of generative models that learn to generate data by reversing a diffusion process. The forward diffusion process gradually adds noise to the data, while the reverse diffusion process attempts to recover the original data from the noisy data.

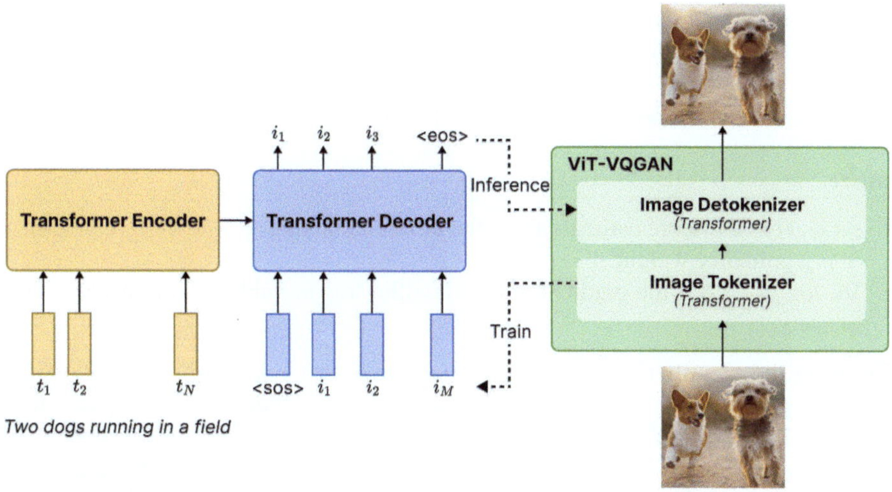

Fig. 1.15 Parti's architecture

The forward diffusion process is typically defined as a series of steps in which Gaussian noise is added to the data:

$$q(x_t|x_{t-1}) = \mathcal{N}(x_t; \sqrt{1-\beta_t}x_{t-1}, \beta_t I), \tag{1.2}$$

where:

- x_0 is the original data.
- x_t is the data at step t of the diffusion process.
- β_t is a variance schedule that controls the amount of noise added at each step.
- $\mathcal{N}(\mu, \sigma^2)$ represents a Gaussian distribution with mean μ and variance σ^2.

The reverse process, parameterized by a neural network θ, aims to denoise the data and recover the original data distribution:

$$p_\theta(x_{t-1}|x_t) = \mathcal{N}(x_{t-1}; \mu_\theta(x_t, t), \sigma_\theta(x_t, t)I). \tag{1.3}$$

The model is trained to minimize the difference between the true forward process and the learned reverse process, often using a variational lower bound.

For the conditioned generation, there exist two paradigms for diffusion models to utilize the condition, Classifier-Guided [34, 143] and Classifier-Free [68, 182]. For Classifier-Guided ones, an additional condition classifier is added to a well-trained diffusion model. Classifier-guided diffusion models enhance the text-to-image generation process by incorporating gradients from a pre-trained classifier. This guidance helps steer the generation process toward producing images that are more aligned with the given text prompt. The forward diffusion process and the reverse process are illustrated in Eqs. (1.2) and (1.3). The score estimate is adjusted during sampling in order to apply classifier guidance. The adjusted score incorporates the gradient of the classifier's log probability with respect to the image:

$$\hat{s}_\theta(x_t, t) = s_\theta(x_t, t) + w\nabla_{x_t} \log p(y|x_t), \quad (1.4)$$

where:

- $s_\theta(x_t, t)$ is the score estimate from the diffusion model.
- w is a guidance weight.
- $\nabla_{x_t} \log p(y|x_t)$ is the gradient of the classifier log probability with respect to the image x_t.

By using the classifier's knowledge of the image-text relationship, this classifier guidance guides the diffusion model to produce images that more closely resemble the textual description.

For Classifier-Free ones, condition signals are added during the diffusion model training process. Classifier-free methods achieve great control of details when equipped with sufficient training data and computing power. Formally, classifier-free diffusion models [68] provide an alternative to traditional classifier-guided models by eliminating the need for a separate classifier. Instead, they use a single neural network to parameterize both conditional and unconditional models. This approach simplifies the training process and reduces computational overhead. The forward diffusion process and the reverse process are the same as above. During training, the network is optimized to denoise the data and match the original distribution. The classifier-free guidance modifies the score estimate for conditional sampling without requiring classifier gradients:

$$\hat{s}_\theta(x_t, t) = (1 + w)s_\theta(x_t, t) - ws_\theta(x_t, \emptyset), \quad (1.5)$$

where w is a guidance weight, $s_\theta(x_t, t)$ is the conditional score, and $s_\theta(x_t, \emptyset)$ is the unconditional score. This method maintains the balance between generating high-quality images and adhering to textual descriptions without the complexity of classifier training.

There are many pioneering works based on DM, including GLIDE [167], Imagen [182], Stable diffusion [180], DALL-E2 [177], latent diffusion model [180]. GLIDE [167] first formalizes the free text-to-image diffusion model. Classifier guidance and classifier-free guidance are introduced, and it demonstrates that classifier-free guidance yields more preferable results in photo-realism and text alignment. It models the input text condition into a token sequence and replaces the class label embedding with the pooled text features in the traditional class-conditioned diffusion models and concatenates the text features to the context in the attention layer. Imagen [182] uses cascaded diffusion models for photorealistic text-to-image generation. It starts with a base model generating a 64 × 64 image, which is then progressively upsampled to 1024 × 1024 using a super-resolution model. The method employs classifier-free guidance to improve text conditioning and incorporates dynamic thresholding to enhance image fidelity and

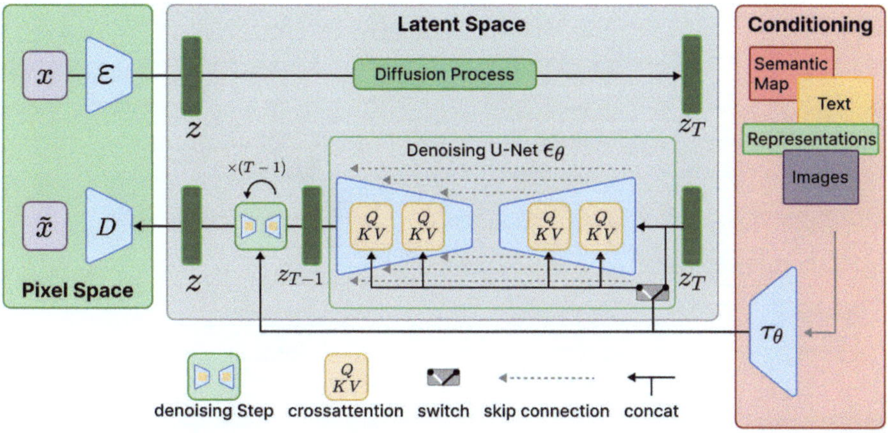

Fig. 1.16 Overview of the Latent Diffusion Model (LDM)

alignment. Noise conditioning augmentation is used in the super-resolution stages to produce high-quality images, and the model is optimized with Efficient U-Net architectures for faster convergence and improved performance. Latent Diffusion Model (LDM) [180] (Fig. 1.16) trains diffusion models in the latent space of a pre-trained autoencoder. It reduces computational complexity while maintaining high visual fidelity by focusing on the perceptually significant aspects of the data. The architecture incorporates cross-attention layers, enabling flexible conditioning inputs such as text or bounding boxes. With substantially lower computational costs than conventional pixel-based diffusion models, LDM exhibits state-of-the-art performance in tasks like image inpainting, class-conditional image synthesis, text-to-image synthesis, and super-resolution.

1.2.3 Video-Text Understanding

There are many differences between videos and images. First, videos are more than just a continuous series of pictures; they typically include motion features, audio, and captions that frequently enhance one another. It makes video understanding a more challenging multi-modal task. Second, for its sequential attributes, videos contain both low-level coherence across neighboring frames and high-level temporal semantic correlations, which should be considered in video modeling. Similarly, this part introduces the video-based task-specific vision-language learning by the following representative tasks.

Video Question Answering (VideoQA) The goal of VideoQA [96, 110, 200] is to answer the questions either in the form of free text in natural language or by selecting one answer out of a set of candidate choices. Compared to images,

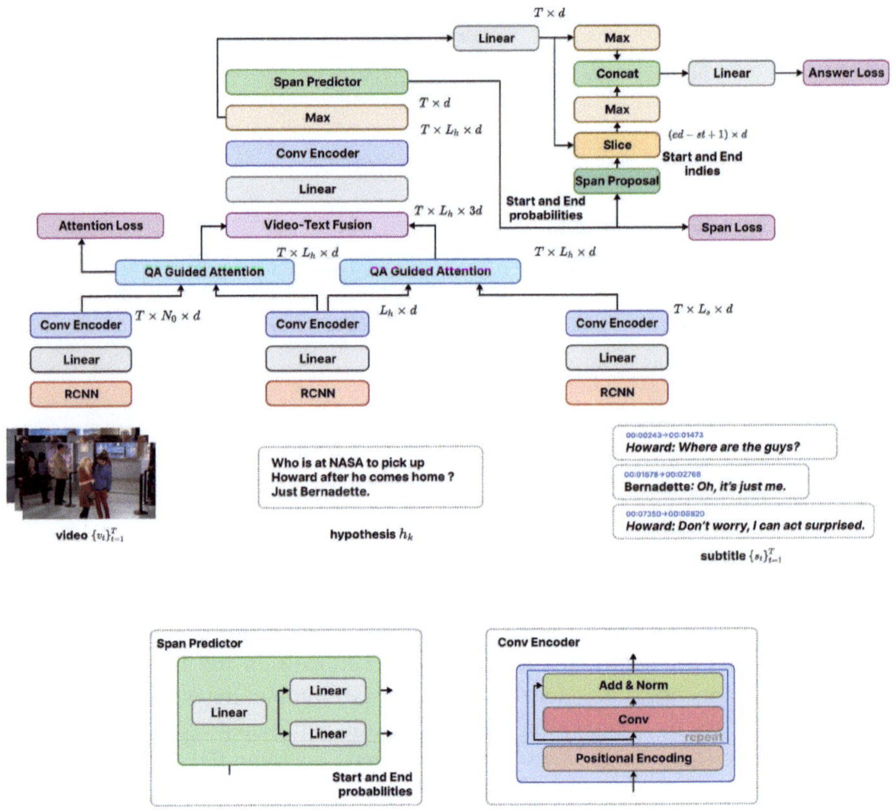

Fig. 1.17 Overview of the proposed STAGE [111] framework

videos contain less information than images, and it is not easy to specifically focus on relevant moments and ideas for questions. Thus, the coarse-to-fine strategy is effective. As shown in Fig. 1.17, Lei et al. [111] augment TVQA datasets with bounding boxes linking depicted objects in questions and answers to visual concepts and propose the STAGE framework that grounds evidence in both spatial and temporal domains to answer questions about videos. Shen et al. [185] propose a text-guided object detector, which takes text question-answer pairs and video frames as inputs, to detect the objects relevant to the given text, and thus provide intuitive visualization and interpretable results.

Manual annotation of questions and answers for videos is often tedious and prohibits scalability. Thus, it motivates the development of few/zero-shot VideoQA approaches [227, 228, 240]. Yang et al. [230] enable zero-shot VideoQA through masked language modeling using frozen bidirectional language models.

Since videos contain more objects and much more complex plots to be dug out, a more challenging extended form of VideoQA is the video-centric dialogue system (VideoDial) [3, 71], which requires the model to reason based on as much information as possible in videos and dialogue histories in multiple rounds of QAs. Early works [71, 105, 166] use the recurrent neural network to encode dialog history. Besides, transformer-based VideoDial methods emerge with query-aware attention [106], word-embedding attention decoder [109], and pointer-augmented decoder [104] techniques. Geng et al. [61] generate scene graphs within each frame and aggregate them over the temporal axis to model fine-grained information flow in videos. Kim et al. [98] propose a Structured Co-reference Graph Attention mechanism for a video-grounded dialogue system, which decodes the answer sequence to a question about a given video while maintaining the dialogue context. They use a Structured Co-reference Resolver that performs dereferencing by constructing a structured graph across multiple modalities, and a Spatio-temporal Video Reasoner that captures the local-to-global video dynamics through progressively neighboring graph attention.

Expressive features are important for video-based multi-modal tasks. One of the differences of video-based and image-based vision-language learning lies in the visual features. End-to-end methods are usually used in image-based tasks, while pre-extracted features are used in video-based tasks because of the larges scale of video data. Some works try to extract better video features in an end-to-end fashion to improve the expressiveness of features.

1.2.4 Text-to-Video Generation

Text-to-video generation has garnered widespread attention. Video generation involves modeling temporal dynamics to produce coherent and realistic content, which requires significant computational resources. Additionally, models must often adhere to principles such as physics, user visual comfort, and logical content composition when generating video clips. Furthermore, the scarcity of high-quality training data makes video generation even more challenging than image generation.

Similar to text-to-image generation, text-to-video generation also employs generative machine learning techniques, such as VQ-VAE, GANs, autoregressive models, and diffusion models, which are similar to those used in image generation. In order to ensure that the output content corresponds with the intended distribution implied by the text instructions, user-provided text prompts are used as a condition for generation [26].

The main challenge in text-to-video generation is to process temporal information. As an important construction block in this area, a temporal handler is a critical element that complements the vision processor, which learns the dynamics in frame progression. Figure 1.18 illustrates the main types of temporal handlers, as detailed below.

Temporal Attention Adding a temporal layer to text-to-vision generation tasks is a direct way to incorporate temporal information, especially in autoregressive models such as those used in text-to-video generation. This can be accomplished

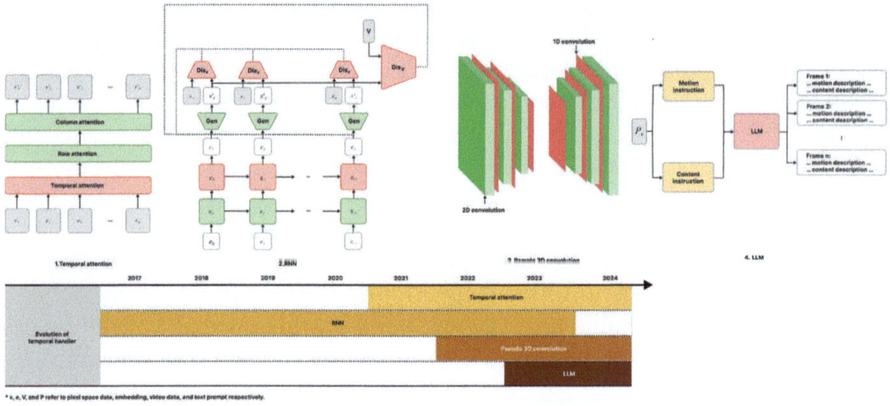

Fig. 1.18 Types of temporal handler in text-to-video generation

through several methods, including the use of an axial transformer [74] to address the temporal dimension, spatio-temporal attention [66] mechanisms to focus on both spatial and temporal features, and causal transformers [80] in the encoder module to preserve the sequence's causality. Additionally, more sophisticated approaches are also used [1], such as the neural Ordinary Differential Equation (ODE) [144] that approximates temporal dynamics and the bidirectional masked attention transformer. These methods rely on the tokenization of input frames during training and the incorporation of conditions such as context memory and motion anchor to ensure the consistency and diversity of the generated frames.

RNN For text-to-video generation models lacking an autoregressive architecture, such as GAN models, incorporating a recurrent neural network (RNN) is a prevalent method to manage the temporal dimension. Commonly-utilized RNNs include Long Short-Term Memory (LSTM) [70] and Gated Recurrent Unit (GRU) [25], which generate temporal sequences from the input text prompt. The RNN processes the encoded text representation t_f, noise $z_f \sim \mathcal{N}(0, I)$, and its previous hidden state h_{f-1} to produce the hidden state h_f, which is then fed into the generator to produce the video frame at time f. To achieve both temporal consistency and content diversity, GAN-based text-to-video generation models often integrate an additional module that acts as a hint generator, connecting the generated content across frames. This module can be implemented in various ways, such as through an additional frame discriminator [97], a gist layer that blends and dynamically updates local and global context information, and Memory Augmented Recurrent Transformer (MART) [112] that memorizes the content described in the global narrative while decomposing it into frame-level captions [155].

Pseudo-3D Convolutions and Attention By appending 1D convolutions post 2D convolutions, models achieve efficiency and preserve spatial knowledge from pre-trained networks while learning temporal features [27]. The pseudo-3D attention,

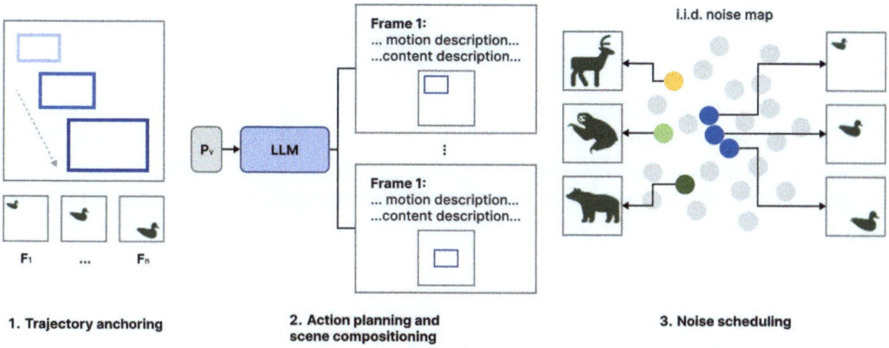

Fig. 1.19 Anchoring trajectory method that injects temporal motion conditions

inspired by VDM [69], further enhances this by sharing temporal insights across the input. Make-A-Video [189] popularizes these techniques, which are now standard in adapting DDPM for video generation. To ensure temporal coherence and diversity, DDPM models often integrate these with noise scheduling, alignment, and other consistency techniques.

Large Language Model Recent advancements in text-to-video generation leverage Large Language Models (LLMs) for their proficiency in multi-modal and multitask learning. Models proposed by Huang et al. [81] and Lu et al. [150] utilize LLMs to transform simple text instructions into detailed scene descriptions, which are then directly fed into the text-to-image generation module. Alternatively, some models [47, 126] adopt a nuanced approach, employing LLMs as temporal encoders to produce scene information that enhances the step size and main text prompt, similar to adding motion anchors to the generation pipeline.

How to integrate motion conditions into the video generation process is important in the text-to-video area. Sometimes, more high- and low-level signals are also exercisable. Anchoring trajectory (Fig. 1.19) is considered the most apparent method in providing inter-frame temporal guidance for text-to-video models, which often embeds physical signals in the pixel space to represent particular motions of the foreground objects, such as bounding box changing positions [84], segmentation mask [94], depth map [21], optical flow [127] and scribble [35].

Besides, another way to control temporal motions is by utilizing Large Language Models (LLMs) to convert simple captions into detailed scene compositions. LLMs can expand prompts in multiple sequences [132], potentially assisting in coding for physics engine simulators like Blender [153] or sketching spatial signals for object propagation [126].

In diffusion models, noise is a critical element that dictates the generated content. Typically, Gaussian noise is preferred for its ability to produce diverse outputs. However, in the text-to-video generation, simply using independent Gaussian noises can compromise performance due to the need for inter-frame consistency [81]. The noise must be initialized in a way that frames interact through temporal attention, ensuring either spatial clustering or a coherent temporal sequence [11, 60, 175]. Noise scheduling is a burgeoning area of research in video generation. Early methods such as progressive scheduling, which is introduced by PYoCo [60], generate noise for subsequent frames by perturbing the previous frame's noise. More recent approaches include calibrating noise to zero at the terminal step, shuffling frame noises, shifting noise in accordance with motion flow, and joint noise sampling for all frames [21, 63, 81, 175].

1.3 Vision-Language Pre-training

Since a unified framework is missing in task-specific Vision-Language learning, some works scale up the model and design pre-training strategies, making them suitable for multiple downstream tasks. This part will give overviews on Vision-Language pre-training, including frameworks and relevant proxy tasks, especially on video-language pre-training [56]. Relationships between image-language and video-language pre-training will also be introduced.

The main challenge with VL pre-training is comprehending the text, the visual content, and the relationship. A large data source is necessary for this target. For image-language pre-training, large-scale datasets of image-text pairs are the main training data. A unified image-language pre-trained network architecture can be broken down into an image encoder, text encoder, and multi-modal fusion module. The image encoder can be divided into two types: sparse feature encoder (e.g., object detector) and dense feature encoder (e.g., CNN, ViT). Common text encoders contain embedding, transformer-based (e.g., BERT, RoBERT). For multi-modal fusion, Zhou et al. [257] and Chen et al. [15] concatenate two modalities input and adopt a modality-unaware fashion through self-attention based on transformer encoder, while Tan et al. [197] and Li et al. [117] adopt a modality-aware fashion and combine self-attention and cross-attention based on transformer encoder as well. Similar methods are used by others [5, 44, 209, 236] to model multi-modal fusion based on transformer decoders. The main pre-training tasks contain image-text contrastive learning, image-text matching, masked language modeling, language modeling, etc. There are some representative works of image-text pre-training. Lin et al. [133] (Fig. 1.20) innovate in visual language pre-training by introducing VILA and pioneer the use of interleaved visual language data for pre-training, demonstrating its effectiveness in maintaining text-only capabilities and improving task accuracy in visual language tasks.

DreamLIP [256], c.f. Fig. 1.21, uses long captions generated by a pre-trained multi-modal Large Language Model (MLLM) to enhance the semantic richness of language-image pre-training. By dynamically sampling subcaptions and introducing a subcaption-specific grouping loss, DreamLIP achieves superior fine-grained alignment between text and images, outperforming previous methods with fewer training pairs.

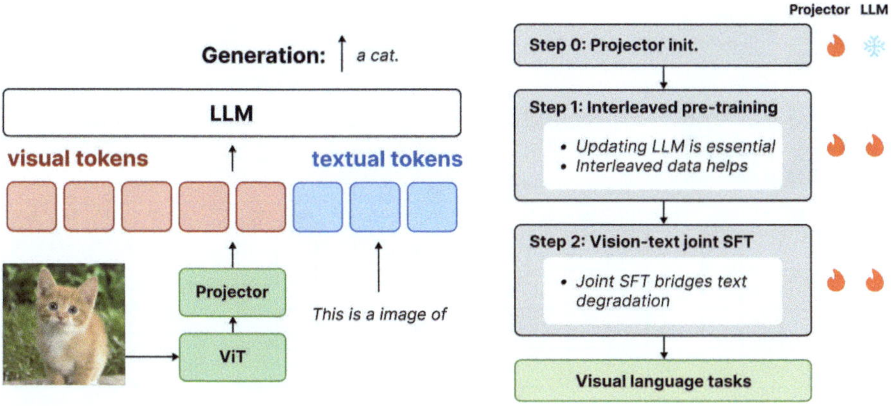

Fig. 1.20 The pre-training pipeline designed by [133]

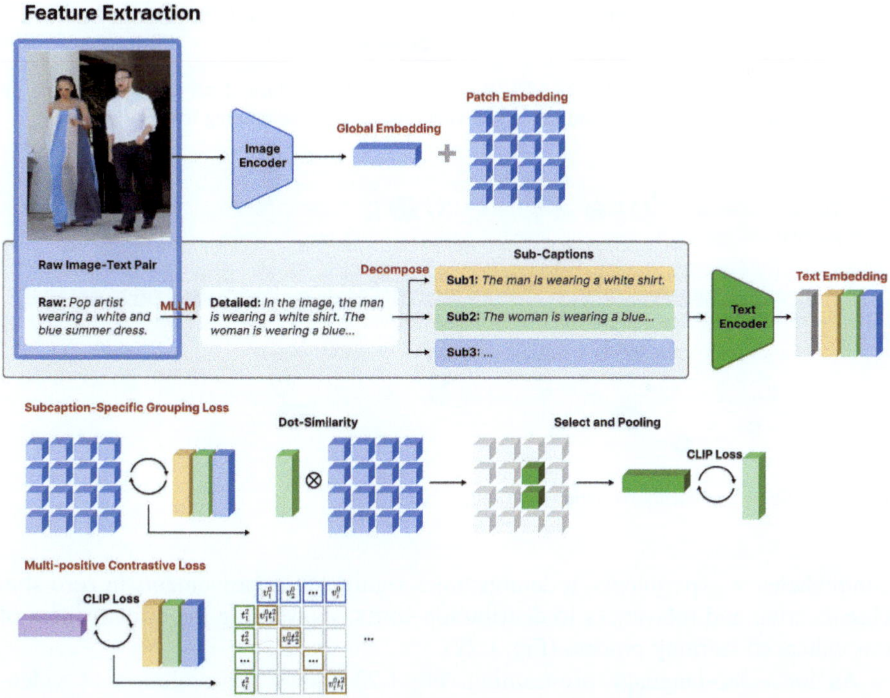

Fig. 1.21 The framework of the DreamLIP [256] pre-training process. For fine-grained local-level text-image alignment, it introduces a grouping loss to match the embeddings of each sub-caption with its corresponding local image patches in a self-supervised manner

RANKCLIP [253] introduces a novel pre-training approach that uses ranking consistency to capture complex many-to-many relationships between text and images, going beyond the strict one-to-one matching framework of CLIP. Through

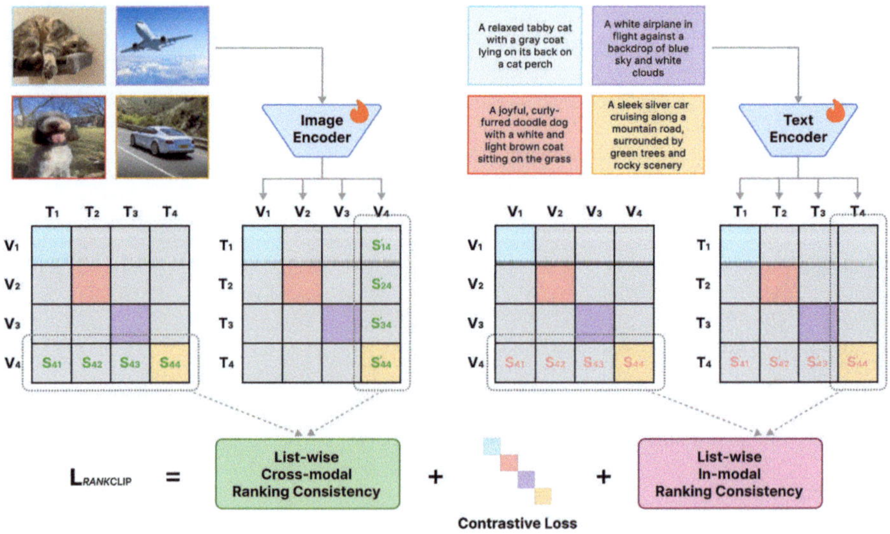

Fig. 1.22 The framework of the RankCLIP pre-training. It introduces both cross-modal and in-modal consistency terms by minimizing a self-supervised, list-wise ranking loss

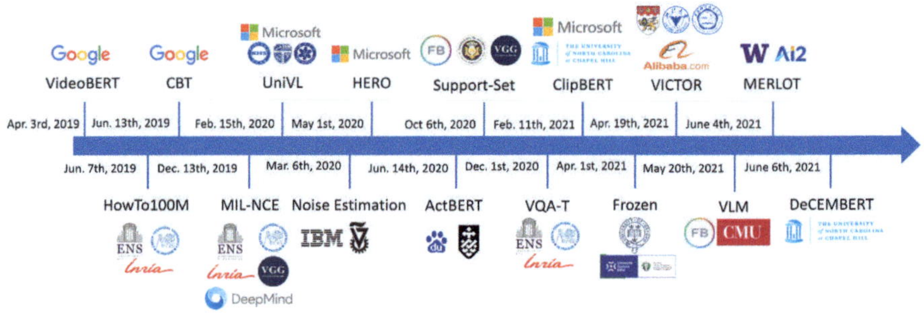

Fig. 1.23 Some representative video-language models [56]

comprehensive experiments, it demonstrates significant improvements in zero-shot classification and robustness to distribution shifts, showcasing the effectiveness of this enhanced learning process (Fig. 1.22).

As for video-language pre-training, Fig. 1.23 shows the evolution of video-language pre-training. Different from images, it's easy to get subtitles or automatic speech recognition transcripts for videos, making "free" annotations accessible for videos, which makes it possible to construct large-scale video-text datasets. Pre-training on large amounts of datasets is helpful for performance improvement on target tasks without heavy post-training, which is non-trivial for videos whose data scale is large due to its information sparsity (Fig. 1.24).

Some pioneering works [151, 161, 194] explore video-language pre-training. VideoBERT [194] (c.f. Fig. 1.25) proposes a joint visual-linguistic model to learn

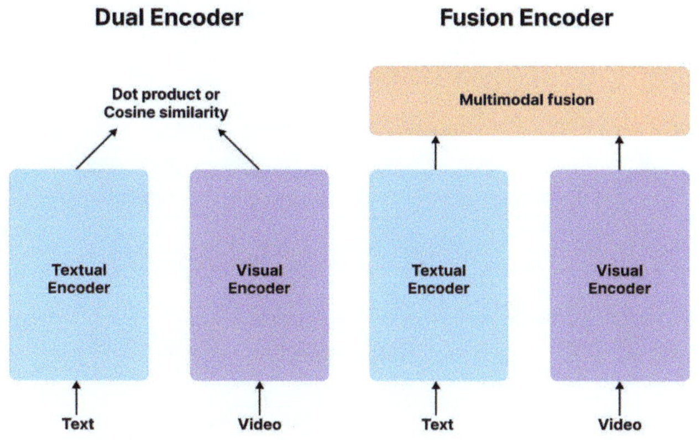

Fig. 1.24 Some representative video-language models [56]

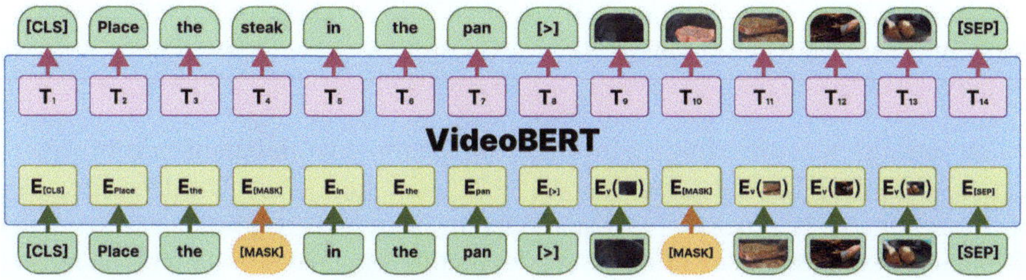

Fig. 1.25 The architecture of VideoBERT [194]

high-level features without any explicit supervision. Specifically, VideoBERT builds upon the BERT model to learn bidirectional joint distributions over sequences of visual and linguistic tokens. UniVL [151] (c.f. Fig. 1.26) is designed for both multi-modal understanding and generation combining five tasks and two training strategies. SRTube [107] (c.f. Fig. 1.27), an innovative video-language pre-training framework that advances cross-modal alignment by integrating action-centric video tube features with semantic role labeling. This approach not only captures the temporal dynamics of video content through the generation of tube features along object trajectories but also enriches language encoding with high-level action-related knowledge. The framework is further enhanced by two novel pre-training objectives and a self-supervision strategy, which collectively contribute to its exceptional performance on various downstream tasks, marking a significant advancement in the domain of video-language pre-training. VLSA [163] (c.f. Fig. 1.28) integrates synchronized audio to learn tri-modal representations through a unified self-supervised transformer, enhancing the joint understanding of visual, textual, and auditory content. VLSA employs local-patch masked modeling to capture modality-aware features and global audio matching to align audio with video and text, resulting in improved performance on cross-modal retrieval tasks with a significantly reduced dataset for training (Fig. 1.29).

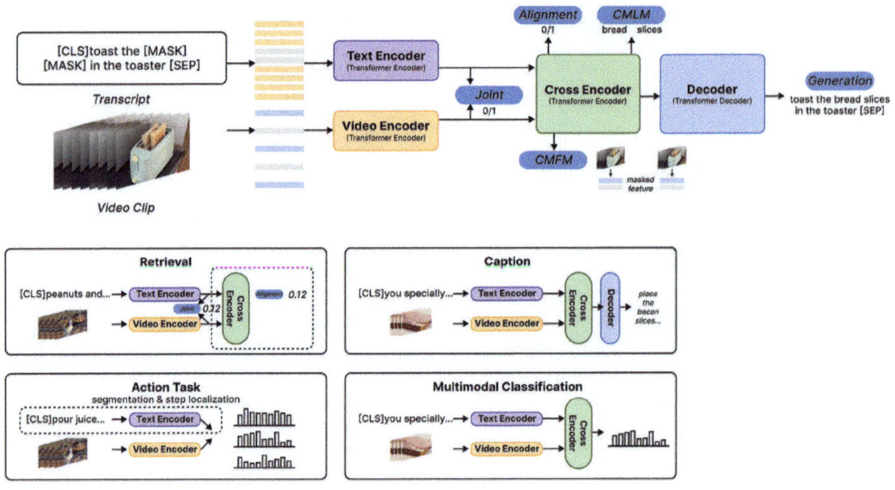

Fig. 1.26 The architecture of UniVL [151]

To sum up, most existing approaches can be roughly classified into two categories: dual encoder and fusion encoder [56] (c.f. Fig. 1.24). The Dual Encoder structure separately processes text and video through their respective encoders and then combines the representations using dot product or cosine similarity. The Fusion Encoder structure processes text and video through their respective encoders and then integrates the representations through a multi-modal fusion layer.

Pre-training tasks are important in video-language pre-training. Video-text contrastive learning aims to learn the correspondence between video and text. MIL-NCE [161] considers a set of multiple positive candidate pairs to mitigate the misalignment between the video and its transcript. Masked language modeling and video-text matching are also common pre-training tasks. All of these tasks, however, only treat the video as a whole without providing oversight of the temporal and spatial relationships within it. Masked video modeling task, similar to masked language modeling, is developed to reconstruct the masked input visual tokens. Since visual features are high-dimensional and continuous, this vanilla masked video modeling exhibits minimal to negligible effects during the pre-training phase. To address this issue, VIOLET [52] (c.f. Fig. 1.30) employs masked visual-token modeling as an improvement of masked video modeling, aiming to reconstruct the discrete latent codes from pre-trained DALL-E. These discrete latent codes work as information compression, decreasing redundancy and dimension, so it achieves promising performance improvements. Frame Order Modeling (FOM) [116, 239] is another task to model the temporality of videos. During training, a percentage of frames are randomly selected to be shuffled and the goal is to reconstruct their original temporal order. FOM can be formulated as a classification task of predicting the timestamp labels. To better boost fine-grained region-entity cross-modality alignment, object-level supervision [118, 208, 258] is proposed. ActBERT [258]

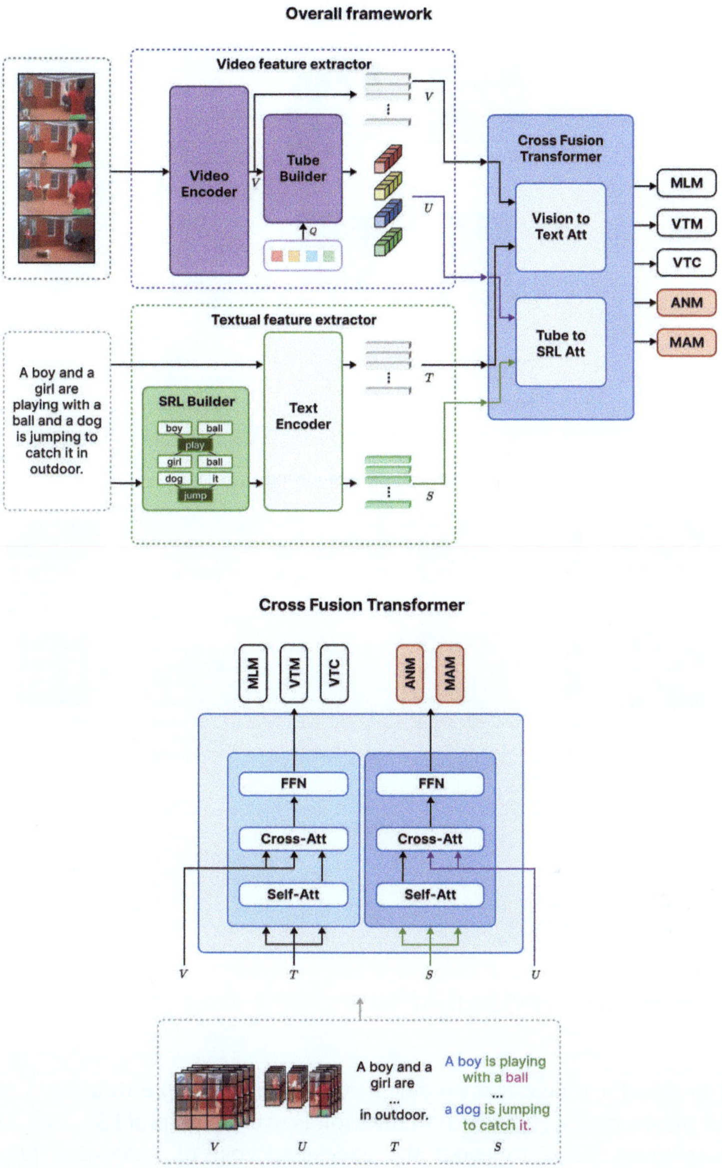

Fig. 1.27 The architecture of SRTube [107]

leverages global action information, linguistic texts, and local regional objects by a unified ActBERT architecture and models object-level supervision by masked object classification task.

There are some special characteristics of video-language pre-training compared to image-language pre-training. Given that videos contain image sequences, it is

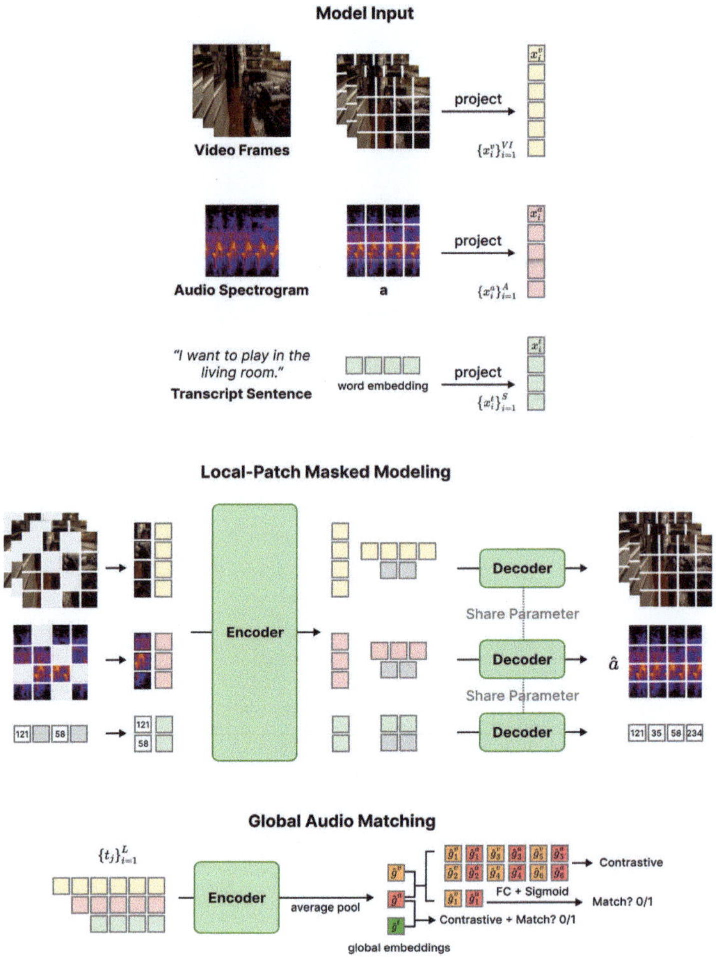

Fig. 1.28 The architecture of VLSA [163]

intuitive to transfer image-text models to video-language pre-training. Leveraging ImageNet pre-trained weights as initialization is usually helpful [24, 140, 210, 245]. Such transferring helps to avoid the excessive cost of video-text pre-training (Fig. 1.31).

Besides joint temporal and spatial relations in videos, another discrepancy is that videos can be modeled in multi-modal settings, including visual frames, subtitles, audio, motions, etc. The fusion of complement modalities usually works positively. MMT [55] introduces seven expert features for videos: OCR, speech, face, audio, scene, appearance and motion. Pre-extracted features from expert models are often well-supervised, thus expressive enough for describing videos. For additional

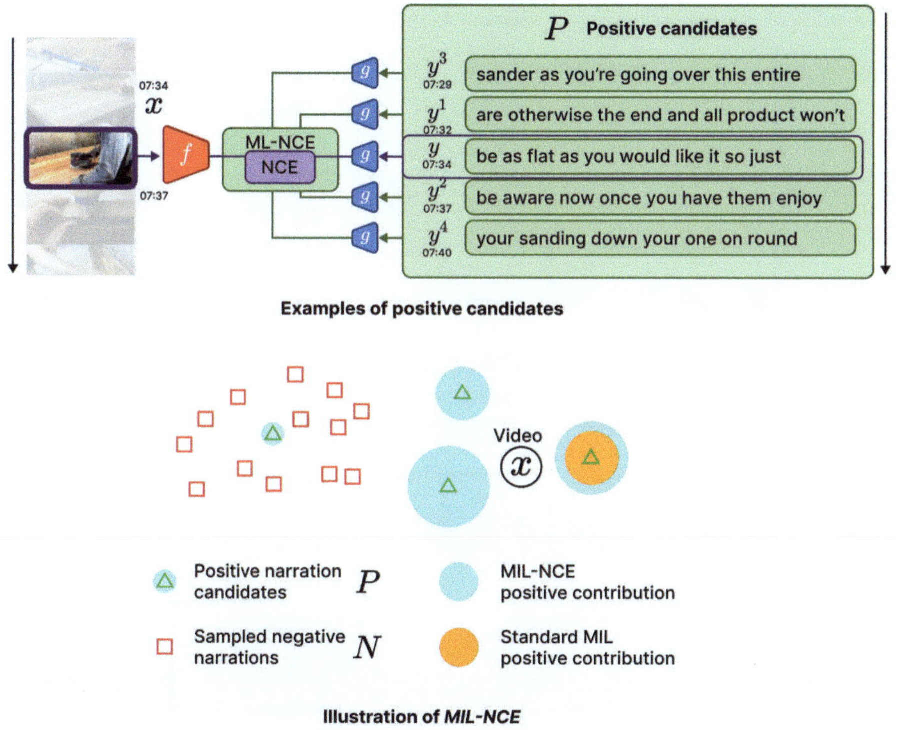

Fig. 1.29 The model architecture of MIL-NCE [161]

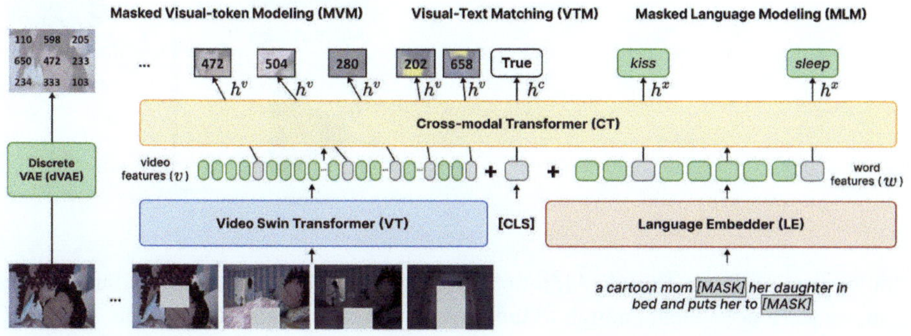

Fig. 1.30 The model architecture of VIOLET [52]

representative approaches that model videos in a multi-modal way, please see additional works [4, 16, 137] (Figs. 1.32 and 1.33).

Compared to images, there are more challenges in video-language pre-training. Noisy transcripts, weak robustness, constrained domains, temporal misalignment, and high computing resources demand make the video language pre-training process

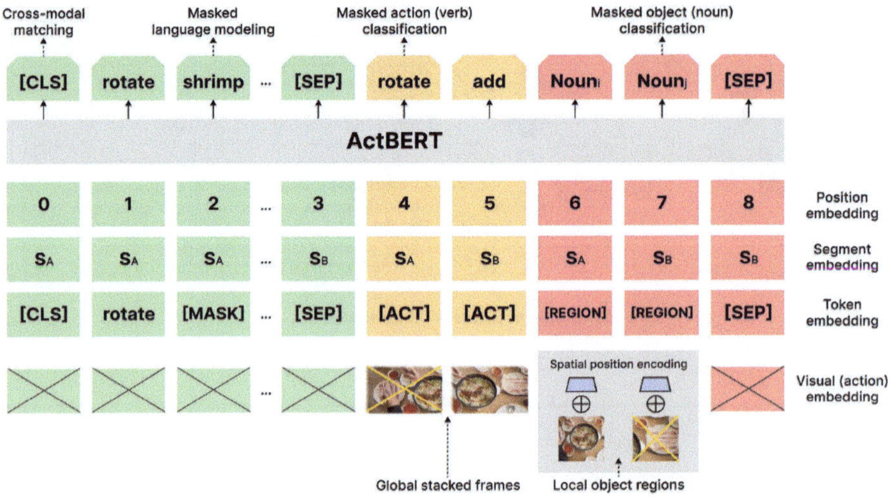

Fig. 1.31 The model architecture of ActBERT [258]

Fig. 1.32 Multi-modal settings of videos

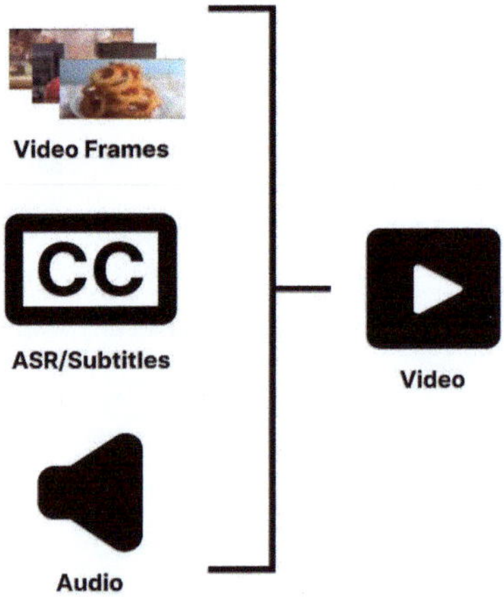

challenging and non-trivial. Although most works are pre-trained on large-scale data, they are not robust enough to the semantic contrast captions, like the temporal order of some details in videos. Bansal et al. [8] (Fig. 1.34) present a dataset that enhances video-language model robustness by incorporating semantically plausible contrast captions generated through large language models, thereby improving performance in tasks such as text-to-video retrieval and video question answering, and demonstrate the effectiveness of this targeted training approach in advancing video-language understanding.

Due to the limited available computation resources, existing video-text pre-training focuses on short videos (less than 16 frames), leaving long videos chal-

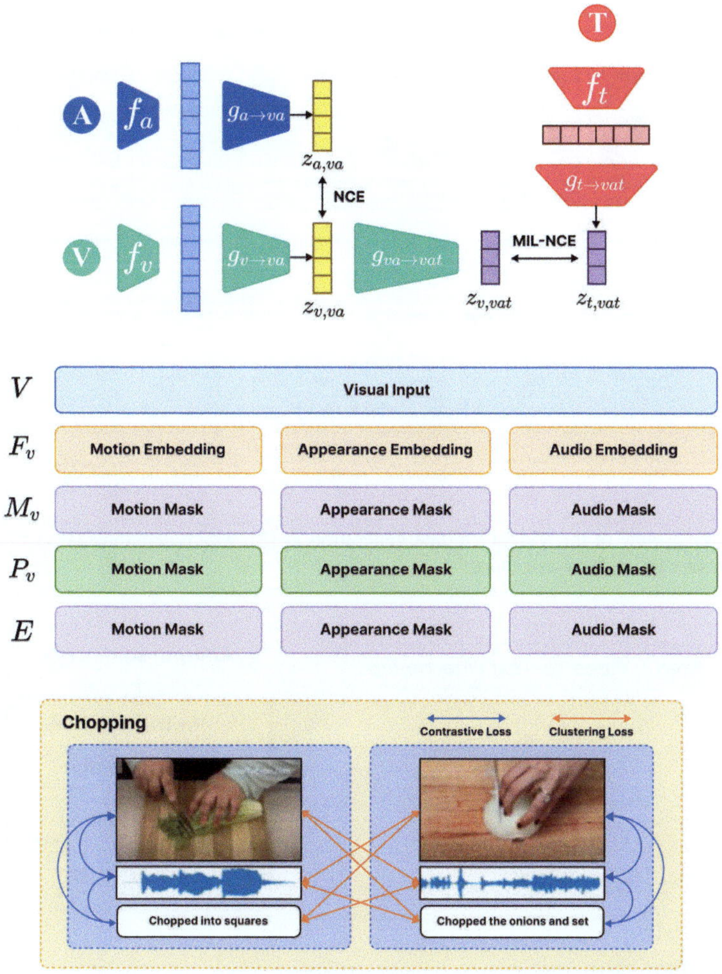

Fig. 1.33 Some representative works modeling videos in a multi-modal manner (Top: [4], Middle: [137], Bottom: [16])

lenging in the video-language area. LONGVIVIT [172] (c.f. Fig. 1.35) presents a breakthrough in scaling video encoders for long-form video understanding, overcoming spatial and memory limitations with a simple yet effective video masking technique during contrastive pre-training. The proposed LONGVIVIT model excels at temporally rich benchmarks, showcasing the advantages of video-first architectures over conventional image-first approaches. This work demonstrates that with minimal architectural complexity, video encoders can be efficiently trained to process videos up to 4.3 minutes, outperforming larger, LLM-assisted models.

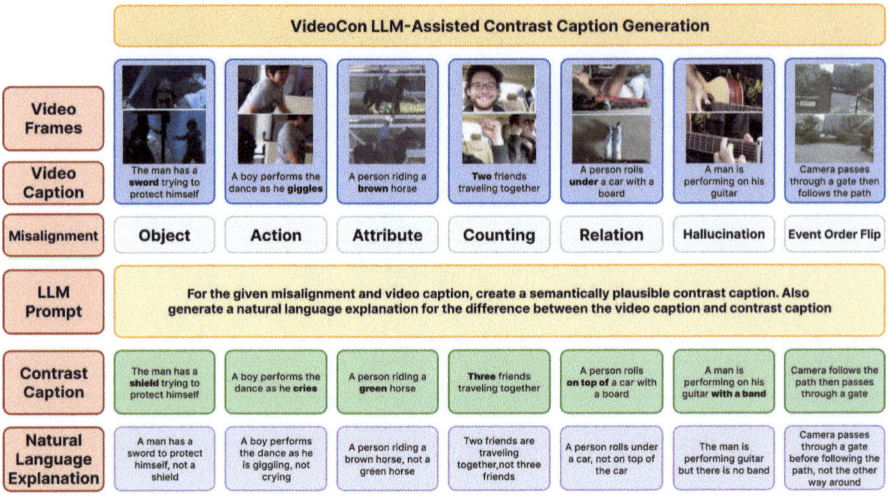

Fig. 1.34 Overview of the VideoCon data generation in [8], which generates types of misalignment contrast caption for pre-training

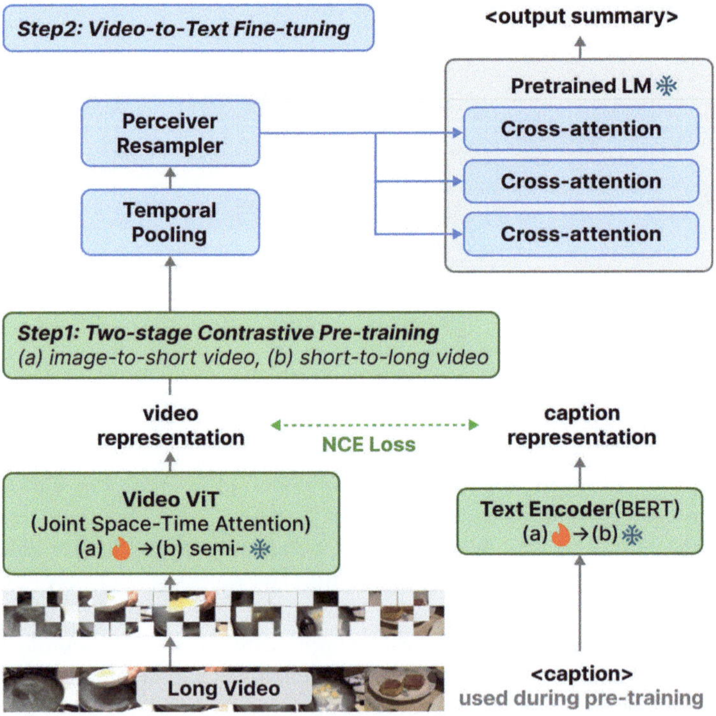

Fig. 1.35 The two pre-training steps in [172]: Two-stage contrastive pre-training and video-to-text fine-tuning

1.4 Multi-Modal LLMs in Vision-Language Learning

In recent days, large language models (LLM) e.g., ChatGPT, ChatGLM, LLaMA spring up, achieving killer performance over a number of benchmarks, especially showing a strong ability on few/zero-shot settings with in-context learning, prompt/instruction learning, etc. Inspired by the huge power of LLMs, many works study multi-modal LLMs which are capable of jointly understanding visual and textual information, such as Flamingo [5], LLaVa [141], GPT-4 [168], VisCPM [202], etc. Some powerful multi-modal generation models are bridged to the LLM and make the MLLM a complete system to understand, process, and generate multi-modal content. The progress of MLLM is shown in Fig. 1.36 [252].

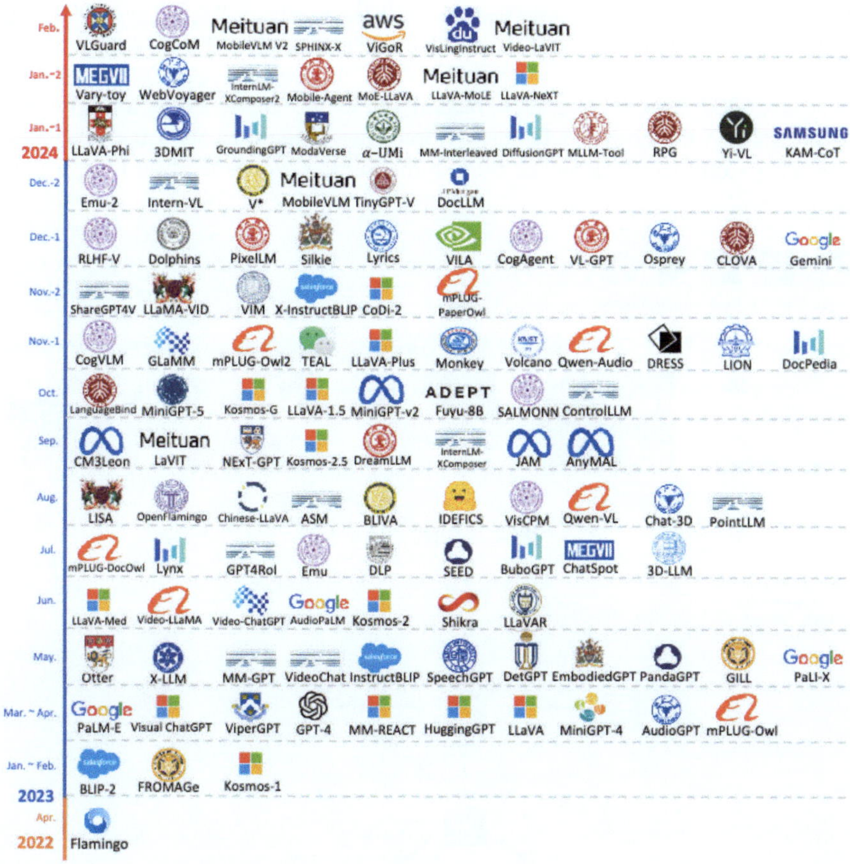

Fig. 1.36 The latest timeline of MLLMs [252]

1.4.1 Basic Modules in Multi-Modal LLMs

The general model architecture of MLLMs can be broken down into five main components: Modality Encoder, Input Projector, LLM Backbone, Output Projector, and Modality Generator. These components work together to process and generate multi-modal content effectively.

The Modality Encoder (ME) is responsible for encoding inputs from various modalities (e.g., image, video, audio) into corresponding feature representations. Different pre-trained encoders can be used for different modalities. For instance, encoders such as NFNet-F6 [12], ViT [38], CLIP ViT, and Eva-CLIP ViT can be used for visual modality, and encoders such as C-Former [17], HuBERT [72], and BEATs are common choices for audio modality. To handle various multi-modal encoders, ImageBind [62] is used as a unified encoder covering six modalities, including image/video, text, audio, heat map, inertial measurement units, and depth (Fig. 1.37).

The encoded features are then passed on to the Input Projector. The Input Projector aligns the encoded features from other modalities with the text feature space. This alignment is achieved by minimizing the text generation loss, which is

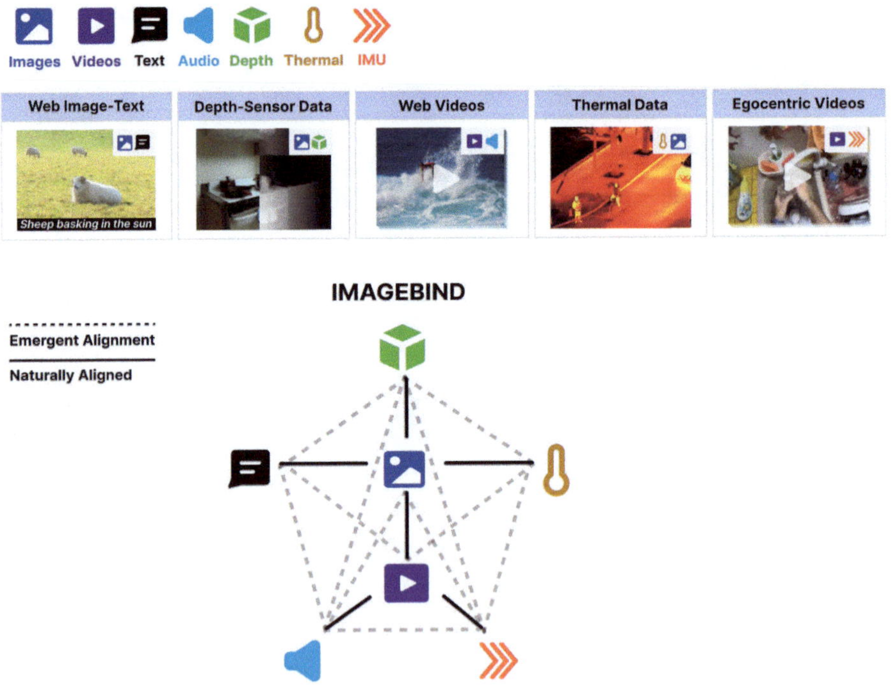

Fig. 1.37 The overview of ImageBind [62]. ImageBind links all these modalities in a common embedding space, enabling the capabilities of emergent alignments

calculated based on the LLM's output and the target texts. Implementations can range from simple Linear Projectors to more complex structures such as cross-attention, Q-Former [121], P-Former [85], and MQ-Former [149].

The LLM Backbone serves as the core of the MLLM, processing representations from various modalities and producing textual outputs or signal tokens for other modalities. Popular LLMs used in MLLMs include Flan-T5 [29], ChatGLM [241], UL2 [201], and LLaMA [204]. These models are chosen for their robust language generation capabilities and zero-shot transfer abilities.

The Output Projector maps the signal tokens from the LLM Backbone into features that can be understood by the Modality Generator. The goal of this mapping is to reduce the distance between the Modality Generator's conditional text representations and the mapped features.

The Modality Generator is aimed at producing outputs in various modalities, such as image, video, or audio. It typically uses off-the-shelf Latent Diffusion Models (LDMs) [192] such as Stable Diffusion for image synthesis or Zeroscope for video synthesis. The features mapped by the Output Projector serve as conditional inputs in the denoising process to generate multi-modal content.

The training pipeline for MLLMs consists of two main stages: multi-modal Pre-Training (MM-PT) and multi-modal Instruction Tuning (MM-IT). During MM-PT, the Input and Output Projectors are trained for alignment among modalities. In the MM-IT stage, pre-trained MLLMs are fine-tuned using instruction-formatted datasets to enhance their zero-shot performance.

1.4.2 Representative Models

From Image-LLMs to Video-LLMs

BLIP-2 and LLaVA represent two distinct architectures in Image LLMs. Specifically, BLIP-2 [121] employs a learnable query transformer as a bridge between vision and language, while LLaVA [142] utilizes a more lightweight linear layer to adapt visual tokens into the LLM's space. As shown in Fig. 1.38, the BLIP-2 model is trained in two stages. In the first stage, it learns the Q-Former, which generates Query Features that possess multi-modal representation capabilities. In the second stage, the model trains the LLM, equipping it with image-to-text capabilities. This two-stage training allows the model to adapt to various downstream tasks, such as Video Question Answering (VQA), Image Captioning (IC), and Image-Text Retrieval (ITR). Building upon the BLIP-2 model, the InstructBLIP model [31] further incorporates text as an input to the Q-Former, enabling the generation of language-aware visual embeddings. This enhancement allows the model to integrate textual information directly into the process of visual representation learning, thereby improving its capacity for multi-modal understanding. Compared to the BLIP series models, which use a more parameter-heavy query transformer as the adapter, LLaVA-like models—such as LLaVA, MiniGPT-4 [260], and MiniGPT-v2 [20]—employ linear layers or MLPs as adapters (c.f., Fig. 1.39). This choice

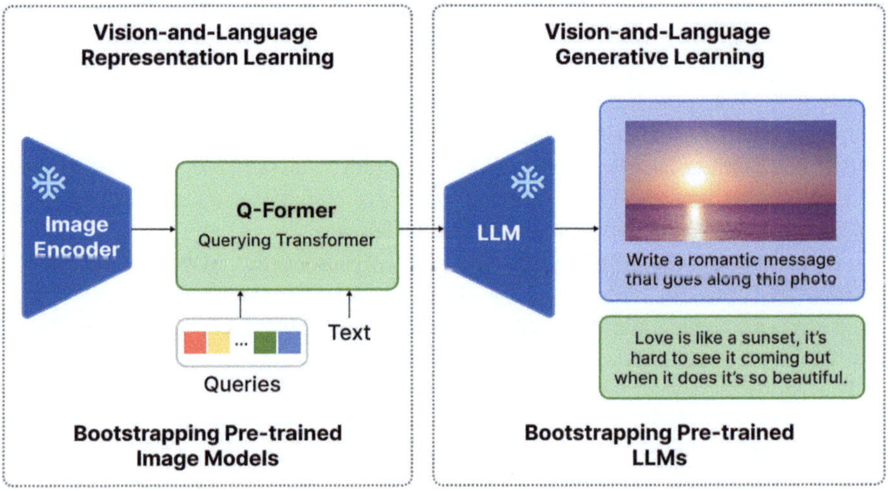

Fig. 1.38 The architecture of BLIP-2 [121]

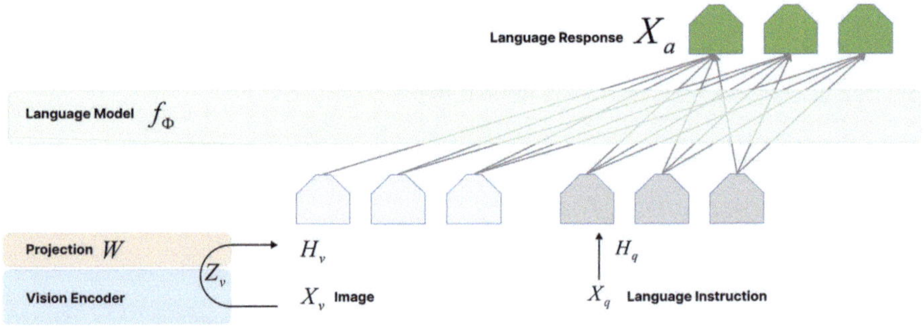

Fig. 1.39 The architecture of LLaVA [142]

results in a significantly lower number of trainable parameters, making these models more lightweight and computationally efficient.

Subsequently, numerous works extend visual signals to the video modality. For instance, VideoChatGPT [154] designs two pooling layers to compress spatial and temporal dimensions separately and develops a high-quality video instruction dataset. Video-LLaVA [131] adopts a pre-trained LanguageBind model [259] to pre-align image and video inputs, as shown in Fig. 1.40. Existing methods typically handle video processing by using representative queries or temporal compression. However, when dealing with long videos, the former approach results in excessively long token sequences, while the latter suffers from significant information loss. Therefore, LLaMA-VID [122] introduces a dual-token approach, compressing each video frame into two types of tokens, i.e., context and content tokens, respectively (c.f., Fig. 1.41).

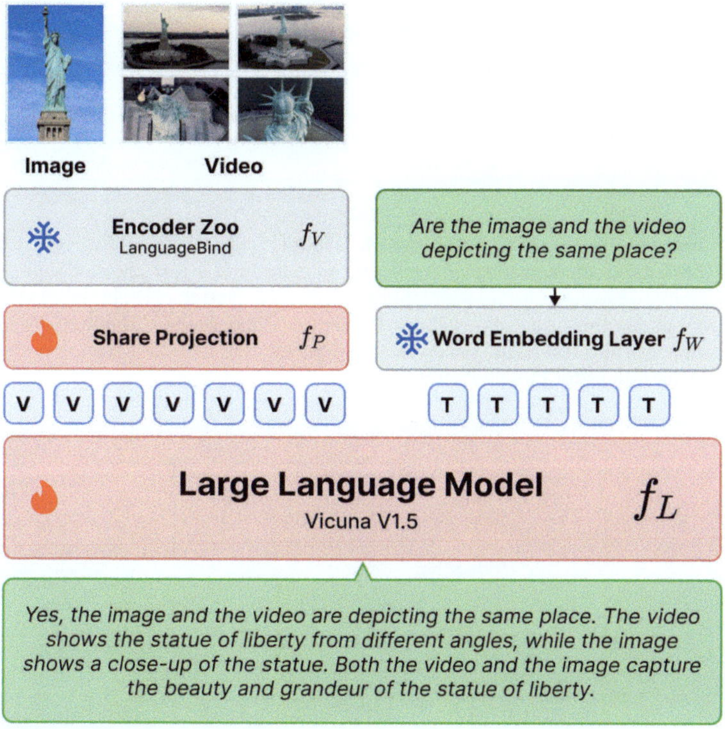

Fig. 1.40 The architecture of Video-LLaVA [131]

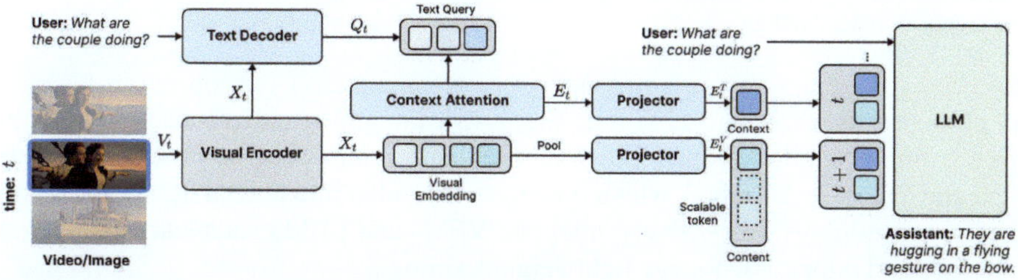

Fig. 1.41 The architecture of LLaMA-VID [122]

LLM can be used as an automated high-quality annotator. Zhao et al. [255] introduce LaViLa, shown in Fig. 1.42, which treats LLMs to be visually-conditioned narrators and fine-tunes over all accessible paired video-text clips. Upon being trained, LaViLa densely annotates videos by generating rich textual descriptions and uses those narrations to train strong dual-encoder models. In short, LaViLa is able to leverage dense, diverse, and well-aligned text generated by an LLM.

Rather than automatically annotating video clips, some video-centric multi-modal systems try to textualize the whole video content. The concept of VideoChat,

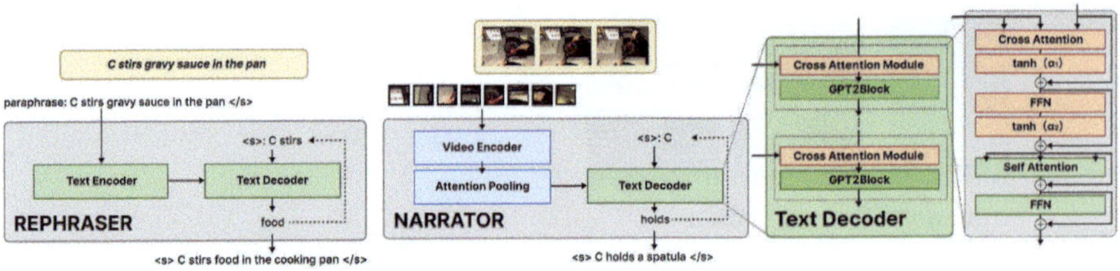

Fig. 1.42 Overview of LaViLa [255], which exploits supervision from two LLMs: (1) Rephraser (Left) takes the narration as input, passes it through a text encoder, and uses a text decoder to autoregressively generate the rephrased output. (2) Narrator (Right) takes video frames as input and obtains the visual embeddings through a video encoder followed by attentional pooling. Equipped with a few additional cross-attention modules, the text decoder autoregressively generates new narrations for those new frames

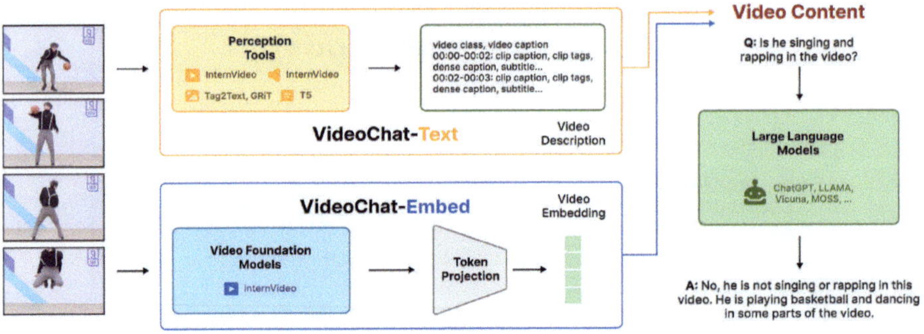

Fig. 1.43 Overview of VideoChat. VideoChat-Text textualizes video in stream, and VideoChat-Embed encodes videos as embedding. Both video contents are input to the following LLMs for multi-modal understanding

as proposed by Li et al. [123] (c.f. Fig. 1.43), integrates video-related tasks into the creation of multiple-round video question answering. They propose VideoChat-Text and VideoChat-Embed, which transform the video into textual descriptions and feature embeddings, respectively, integrate VFMs and LLMs via a learnable neural interface and adopt a two-stage lightweight training.

X-LLM [17], as shown in Fig. 1.44, leverages the advantages of LLMs by treating multiple modalities as foreign languages. Concretely, X-LLM injects multiple modalities, including videos, speech, images, etc., into the LLM through X2L interfaces. This architecture has good scalability to take more modalities into consideration.

Video-LLaMa [249], as shown in Fig. 1.45, takes audio into serious consideration and pays attention to two special challenges in videos, i.e., (i) capturing the temporal changes, and (ii) integrating audio-visual signals. To tackle these two challenges, it proposes a Video Q-former and Audio Q-former. Video Q-former assembles the pre-trained image encoder into a video encoder and learns video-language correspondence through a video-to-text generation task. Audio Q-former is based

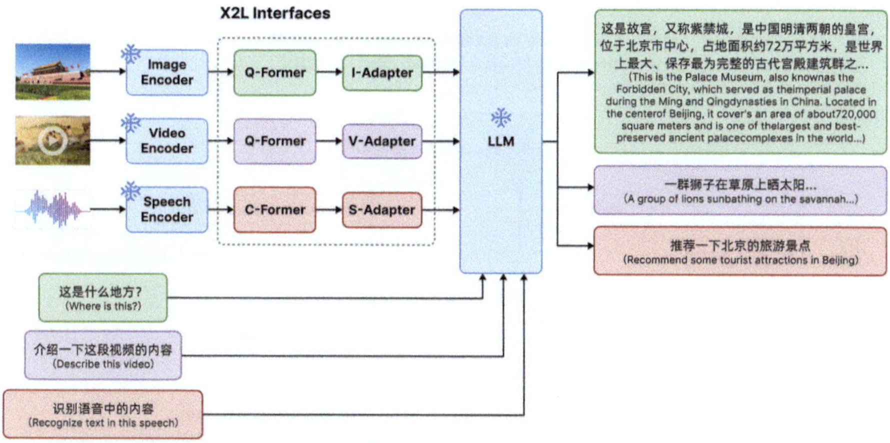

Fig. 1.44 Overview of X-LLM [17]. X-LLM is composed of three training stages: (1) Converting multi-modal Information. Convert multi-modal information into foreign languages through X2L interfaces; only X2L interfaces are updated. (2) Aligning X2L Representations with the LLM. Inject foreign language into LLM, only X2L interfaces are updated. (3) Integrating Multiple Modalities. Integrating multi-modalities, only the adapters in X2L interfaces are updated

on ImageBind [62] to learn reasonable auditory query embeddings for LLM. Then it trains Video-LLaMa on massive video/image-caption pairs as well as visual-instruction-tuning datasets to align the output of both visual and audio encoders with LLM's embedding space.

Besides, readers may refer to other works [152, 154, 211, 215] for advanced approaches promoting video understanding by LLMs.

Video-LLMs for Long Video Understanding

However, such Video-LLMs often lack additional temporal awareness. Temporal modeling, whether at the model or data level, is particularly beneficial for long video understanding. By capturing and leveraging temporal dependencies, models can gain a deeper comprehension of the sequential dynamics in videos, enhancing their ability to process and interpret long-form video content effectively.

MovieChat [193] employs a long-/short-memory mechanism to handle long videos effectively. Initially, it uses the image-based ViT-G/14 of EVA-CLIP and Q-Former to extract visual tokens frame by frame. For short-term memory, it concatenates the features of these frames. When the short memory reaches its limit, the model merges the most similar adjacent frames to form long-term memory. Additionally, MovieChat introduces a benchmark specifically designed for long video understanding, called MovieChat-1K, which is used to quantitatively evaluate the model's performance in this domain (c.f. Fig. 1.46).

In addition, LSTP [212] designs two novel plug-in modules, the Temporal Prompt Sampler and the Spatial Prompt Solver, to compress the number of video tokens. IVA [125] introduces an Interactive Visual Adapter (IVA), which interleaves

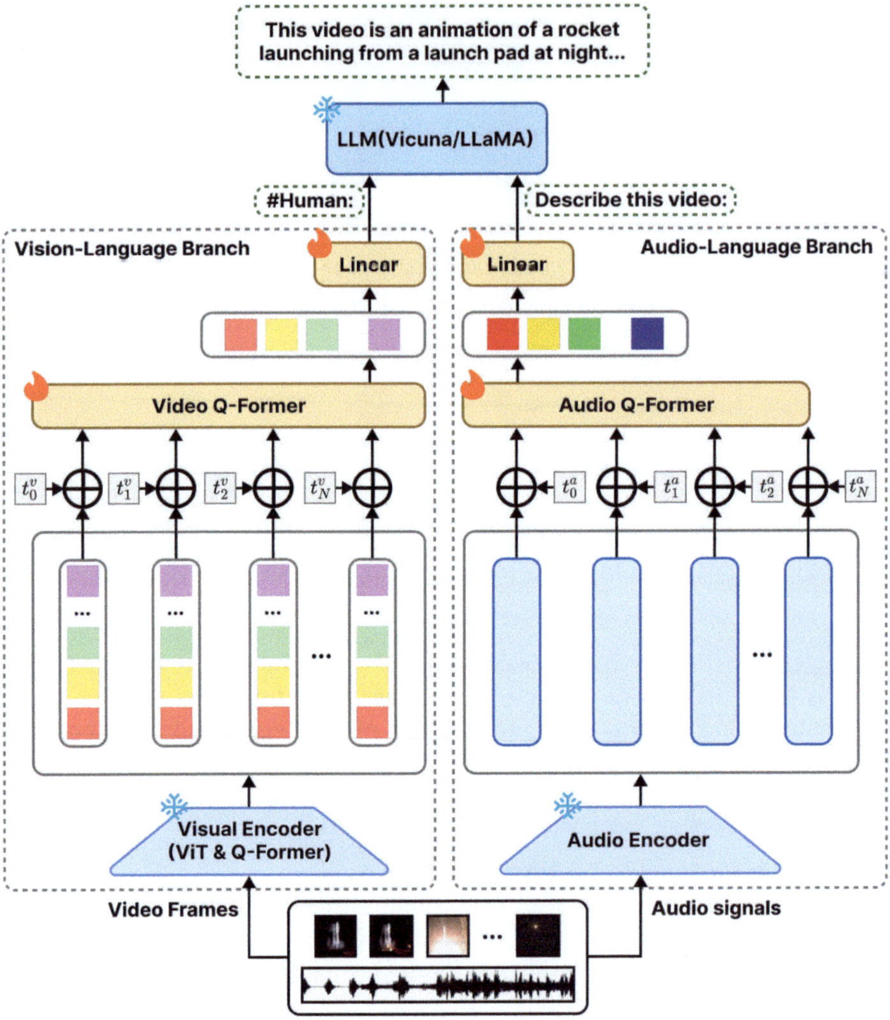

Fig. 1.45 Overview of Video-LLaMa [249], which consists of a two-branch framework to achieve both vision-language alignment and audio-language alignment

video tokens across various layers of the LLM to improve video processing. R-VLM [222] tackles long videos by segmenting them into chunks and selecting the top-k tokens that best match the query to pass on to the LLM. LLoVi [247], a training-free approach, directly uses a pure language-based LLM to answer questions about long videos. Specifically, it divides the video into brief segments, runs each clip through BLIP-2 to produce captions, and then transmits the combined captions and the query to the LLM to get the response.

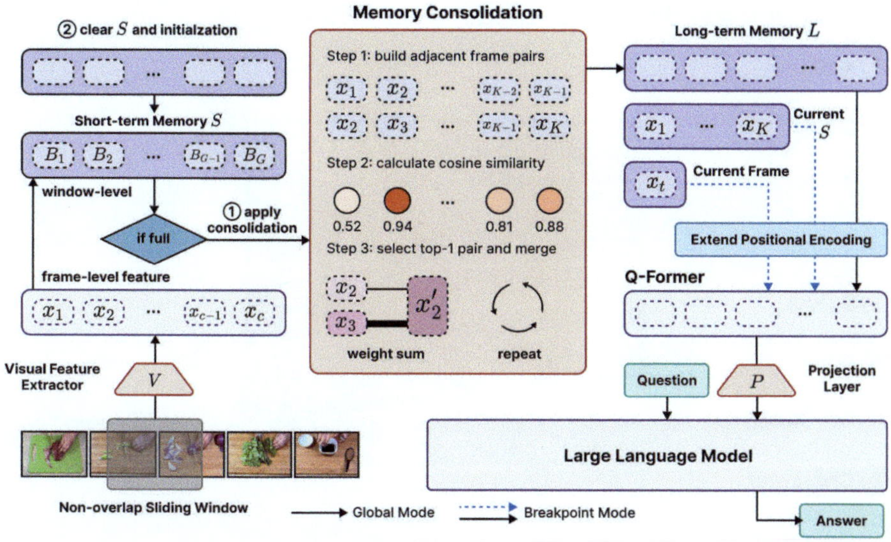

Fig. 1.46 The architecture of MovieChat [193]

1.4.3 Benchmarks for Video LLMs

With the rapid advance in video LLMs, benchmarks are imperative for evaluating the ability of models. There are several benchmarks for video LLM evaluation.

MVBench [124] is a comprehensive benchmark designed to evaluate the temporal understanding capabilities of MLLMs across 20 challenging video tasks. It addresses the gap in existing benchmarks, which primarily focus on spatial understanding in static images, by introducing dynamic tasks that require reasoning over the entire video. Furthermore, the authors present VideoChat2 [124], a robust video MLLM baseline developed through progressive multi-modal training, which significantly outperforms existing models on the MVBench by over 15%. The authors also provide a thorough analysis to guide the design of future MLLMs for enhanced temporal understanding, ensuring the public availability of models and data to foster research in general video understanding (Fig. 1.47).

MMBench-Video [45] (c.f. Fig. 1.48) is designed to evaluate the capabilities of Large Vision-Language Models (LVLMs) in understanding long-form, multi-shot videos. It incorporates a diverse range of video topics and fine-grained capabilities, with an emphasis on temporal reasoning skills. The benchmark has been used to identify significant performance limitations among existing video LLMs, highlighting areas for future improvement in spatial and temporal understanding. Additionally, the study also shows that GPT-4 performs better than other models as an automated assessment tool, providing increased accuracy and robustness.

Video-MME [53] features a diverse dataset of 900 videos with varying durations and modalities, including subtitles and audio, annotated with high-quality

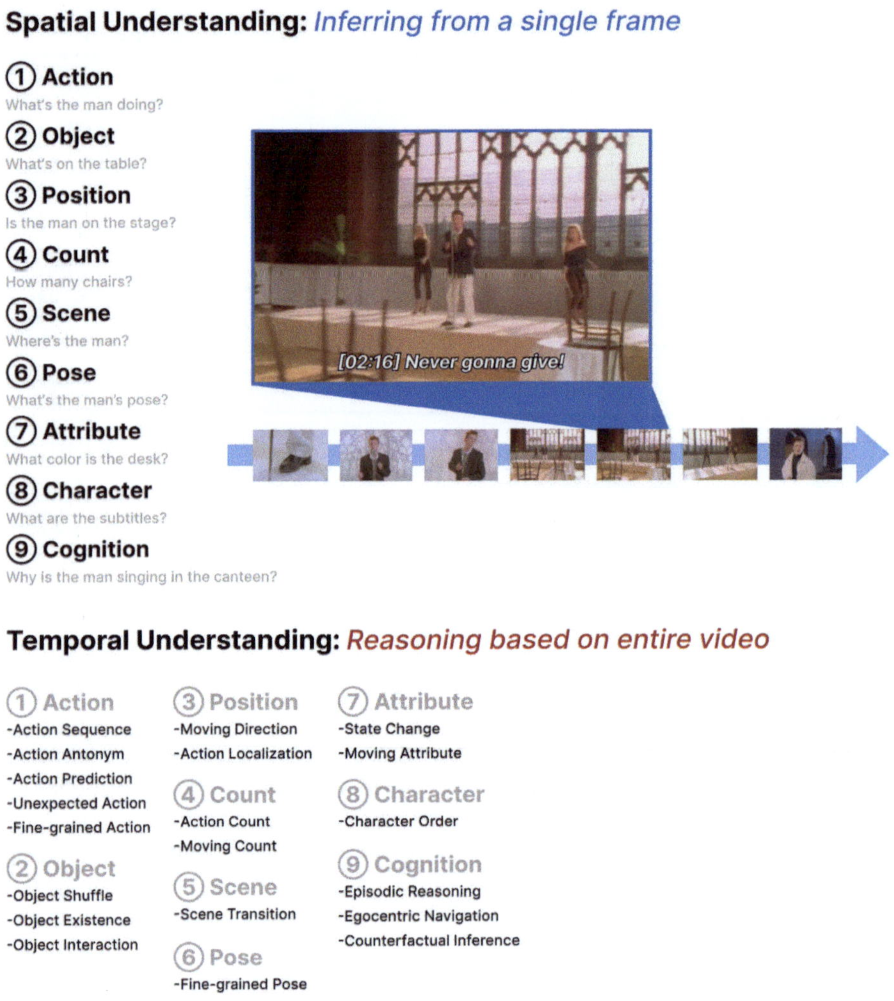

Fig. 1.47 Tasks in the MVBench benchmark

question-answer pairs. The study reveals that commercial models like Gemini 1.5 Pro outperform open-source models, and the integration of subtitles and audio significantly enhances video understanding capabilities.

1.4.4 Existing Challenges in Video LLMs

Though the development of video LLMs is rapid, there still exist several problems in video LLMs, leaving ample room for improvement.

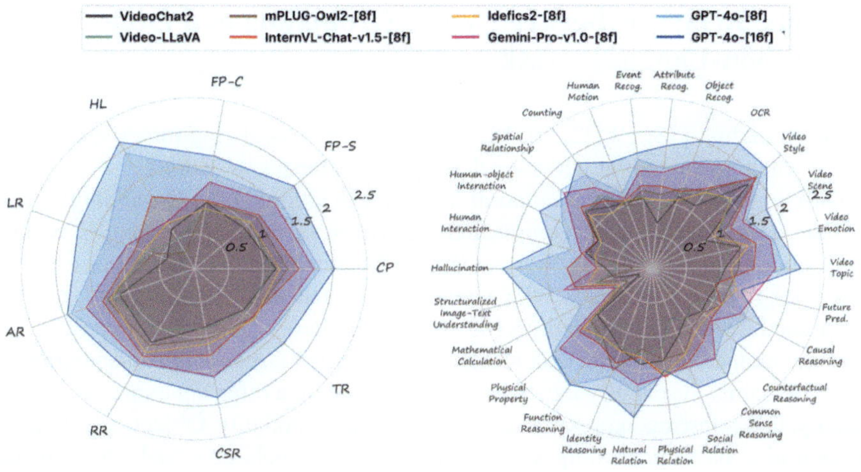

Fig. 1.48 Evaluation of several mainstream methods on MMBench-Video for coarse and fine-grained capability

High Computational Complexity Video MLLMs must process extensive sequences of video frames, which results in a high computational burden. The quadratic increase in self-attention operations with the number of tokens poses a significant challenge for real-time video analysis applications. These models often require substantial GPU resources, making them difficult to deploy on less powerful or cost-sensitive devices.

Data Acquisition and Annotation The need for large-scale, diverse, and high-quality video datasets for training MLLMs is a considerable obstacle. Video data is inherently more complex to annotate compared to static images, as it involves temporal dynamics and contextual understanding. Creating such datasets is time-consuming, expensive, and often requires domain expertise, limiting the generalizability of the models.

Model Generalization and Transferability A critical challenge is the ability of video MLLMs to generalize across different domains and contexts. Models trained on one type of video content may not perform well on another due to variations in video style, language, or cultural nuances. This limits the applicability of MLLMs in diverse real-world scenarios and necessitates continuous research into improving their transferability.

Privacy and Ethical Implications With the rise of video MLLMs, there is an increasing concern about privacy and the ethical use of video data. The models' ability to analyze and interpret visual content raises questions about surveillance, consent, and data protection. It is essential to develop privacy-preserving techniques and establish ethical guidelines for the deployment of these models.

1.5 The Relations Between Video Grounding and Other Vision-Language Tasks

Video grounding [18, 19, 22, 50, 58, 67, 82] requires localizing target segments in videos when given natural language descriptive sentences as queries.

Different from the task of temporal action localization, *Video Grounding* is more flexible since it can locate complicated activities via natural languages, without restrictions from predefined action categories. Meanwhile, *Video Grounding* is more challenging since it requires both textual and visual understanding for semantic alignment between two modalities (i.e., text and video).

In the image domain, the task most similar to *Video Grounding* is visual grounding, which aims to spatially localize an object based on a referring expression [229]. Usually, this referring expression is provided by text modality. Object detection can be seen as a basic grounding task guided by class labels or initial object boxes. More generally, visual grounding can be generalized as a grounding task guided by open-ended descriptions, which provide information on objects, positional relations, interactions, etc. An early standard paradigm for this task utilizes the pre-extracted object proposals [134, 135, 206, 224], followed by more recent works that propose one-stage approaches without relying on proposals, thus achieving higher efficiency [33, 78, 89, 128].

Cross-modal alignment is the common key in other Vision-Language problems as well as video grounding. Techniques in video grounding can be transferred to many Vision-Language problems and vice versa. Vision-Language pre-training aims to learn the aligned representations for videos and text modalities. On the one hand, pre-trained visual and text features from large Vision-Language pre-training models will boost performance for *Video Grounding*. On the other hand, *Video Grounding* is also a brave new and effective proxy task for pre-training that learns the relationship between moment-level videos and flexible sentences.

Video grounding explicitly provides the matching relations between videos and texts. Thus, video grounding can serve as an intermediate task or step for other Vision-Language tasks, such as video question answering [111, 120], thus improving the model's reasoning ability. Li et al. [120] notice the spurious correlations in VideoQA (c.f. Fig. 1.49) and argue that invariant grounding, which discovers the question-critical scenes, is the key to ruling out those spurious correlations and significantly improves causal reasoning ability.

Video grounding can serve as an intermediate task for various downstream vision-and-language tasks, as well as provide interpretability for various tasks such as video summarization and video captioning via effectively linking the text descriptions to specific moments.However, there still exist challenges deserving further research attention:

1. Both video content and sentence queries are inherently sequences carrying rich semantics. As a result, establishing the relationships between videos and

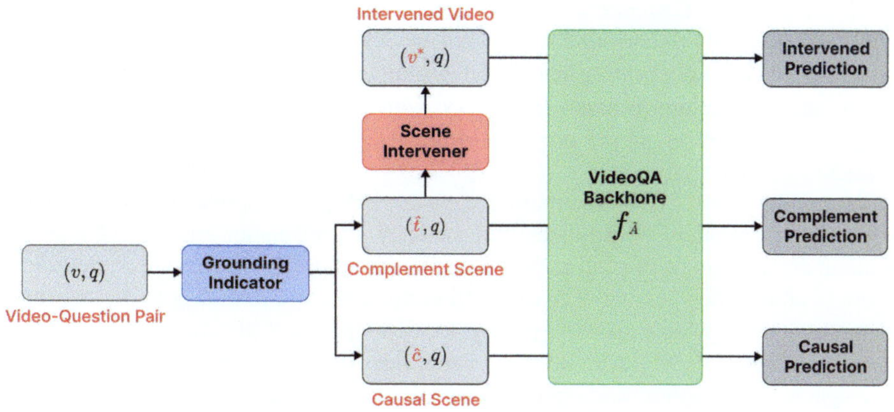

Fig. 1.49 Overview of the IGV model [120], which involves two additional modules, grounding indicator and scene intervener beyond the VideoQA backbone

sentences is a complex task, requiring a fine-grained modeling approach for accurate temporal grounding.

2. The target video segments corresponding to sentence queries vary significantly in both spatial and temporal scales. It would be computationally expensive to extract candidate video segments of varying lengths and locations using sliding windows, and then match them individually with the sentence queries. Thus, efficiently obtaining video segments at different temporal granularities to comprehensively cover the target segments presents a significant challenge for *Video Grounding*.
3. Activities within a video are often not isolated but are semantically interrelated and temporally dependent on one another. Therefore, effectively modeling the video context along with the inherent logical relationships among various video contents, guided by the semantics of the sentence queries, is a crucial yet challenging step for ensuring accurate temporal grounding.

Many promising research works in *Video Grounding* are proposed in order to obtain better solutions to the above challenges, which can be categorized into four groups, ranging from early two-stage matching-based methods, single-stage methods, reinforcement learning-based methods, to weakly-supervised methods. These works will be elaborated on in the Chap. 2.

1.6 Extensions to Generalized Settings

Following a brief overview of the fundamental ideas behind Vision-Language learning tasks, we go on to the possible expansion of the traditional vanilla *Video Grounding* to broader contexts with more difficult problems.

Existing *Video Grounding* datasets follow an implicit assumption that, for a query, there is only one ground-truth moment in a video. As such, one direction for generalized *Video Grounding* lies in the assumption that there could be more than one segment matching a given text query, which is more realistic and reasonable in practice. Lei et al. [113] propose Moment-DETR, an end-to-end transformer encoder-decoder architecture that views *Video Grounding* as a direct set prediction problem. The model takes the extracted video and query representations as inputs and predicts moment coordinates and saliency scores in an end-to-end manner, without any human prior such as proposal generation or non-maximum suppression. From the query perspective, Bao et al. [9] convert *Video Grounding* to dense events grounding task, which aims to jointly localize the multiple moments described in a paragraph, i.e., multiple queries. Then, they devise a DepNet that adaptively aggregates the temporal and semantic information of dense events into a compact set and selectively propagate the aggregated information to every single event with soft attention. Based on DepNet, Shi et al. [186] further present an end-to-end parallel decoding paradigm as language-conditioned regression via employing a transformer-alike architecture from the perspective of *Video Grounding*. Jiang et al. [88] build a graph-based transformer with language reconstruction to jointly extract temporal moments and reconstruct queries with extracted moments for the sake of explainability. Liu et al. [139] also apply a DETR-style framework and introduce pre-training with automatic speech recognition captions for *Video Grounding* task. One inspiration may be that i) localizing multiple moments helps to alleviate the bias in *Video Grounding*, and ii) multi-event joint localization also helps to improve overall accuracy as moments are semantically correlated and temporally coordinated by their order in videos [250].

Another direction for generalized *Video Grounding* may resort to multi-modal settings. For videos, in addition to visual information, some works generalize the traditional *Video Grounding* by using multi-modal information in videos for grounding tasks. Liu et al. [138] use motion and appearance features and model spatio-temporal relations via graph structures. Besides, optical flows and audio information in videos also can provide semantics for videos [14, 19, 138, 139]. For queries, some works focus on more modalities in the queries. In addition to text, additional modalities such as audio, images, and short video clips can also serve as queries for *Video Grounding* [250]. In particular, Audio-Visual Event Localization (AVEL) [39, 203, 213, 220, 223] aims to retrieve the synchronized video segment for a given audio from an untrimmed video. The Image-to-Video Grounding(IVG) task [59, 136, 221, 243] targets at localizing video segments that contain similar activities with those in the query image. Similarly, given a query video and a reference video, Video Re-Localization (VRL) [48, 49, 77, 86] aims to retrieve a segment in the reference video that semantically corresponds to the query video.

Furthermore, it is also possible to generalize *Video Grounding* by adding the space dimension. For example, STGRN [244] proposes Spatio-Temporal Video Grounding (STVG) for Multi-Form Sentences (See Fig. 1.50 for a better understanding of this task). STVG extends the traditional *Video Grounding* in two aspects: (1) localizing spatio-temporal object tubes from untrimmed videos, (2) dealing with multi-form sentences, including both declarative and interrogative sentences. In

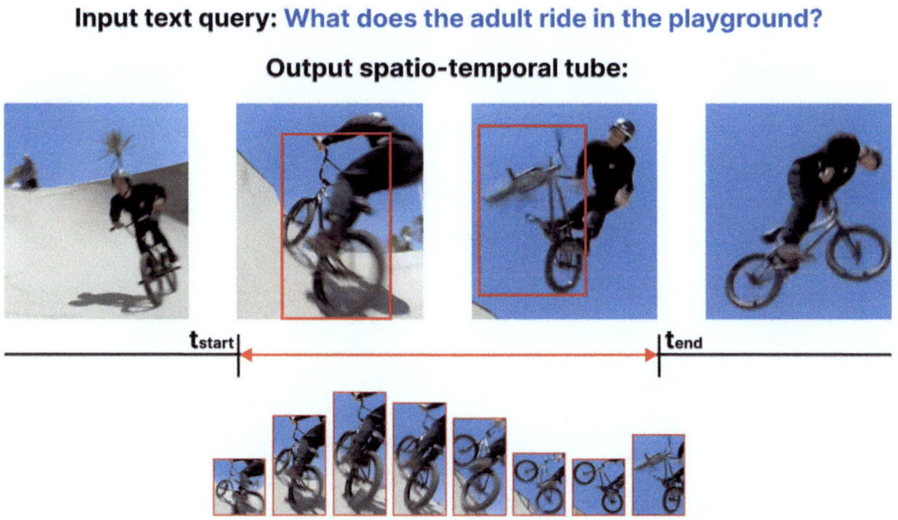

Fig. 1.50 Exhibition of spatio-temporal video grounding [229]

particular, the problem is formulated as a graph problem through building a spatio-temporal region graph to capture the region relationships with temporal object dynamics, which involves the implicit and explicit spatial subgraphs within each frame and a temporal dynamic subgraph across frames. Then textual clues are incorporated into the graph for multi-step cross-modal graph reasoning. Next, a spatio-temporal localizer with a dynamic selection method is introduced to retrieve the spatio-temporal tubes directly without tube pre-generation. TubeDETR [229] also focuses on the STVG task and proposes a novel transformer-based architecture for spatio-temporal video grounding that performs this task with a space-time transformer decoder. It also develops a dual-stream encoder that efficiently encodes spatial and multi-modal interactions based on a slow multi-modal stream and a lightweight fast visual stream (Fig. 1.51).

Given the lack of systematic definitions for these extensions, we systematically formulate them into Generalized *Video Grounding* for the first time in this book, to the best of our knowledge. First, the Video Grounding task is systematically generalized from a comprehensive perspective that encompasses its queries, video content, and the inference process, including specific subproblems such as event grounding and audio-visual temporal grounding. Second, Video Grounding can be conceptualized as a more commonly investigated problem where modalities with richer or higher-density semantics are utilized as queries to ground over modalities with sparser semantic information. Third, Video Grounding also has potential applications in cross-modal retrieval tasks. In summary, given that Video Grounding serves as an intermediate task for video content understanding, it can be further extended by incorporating additional dimensions such as spatial and event-based analysis. More details on generalized video grounding will be elaborated in Chap. 3.

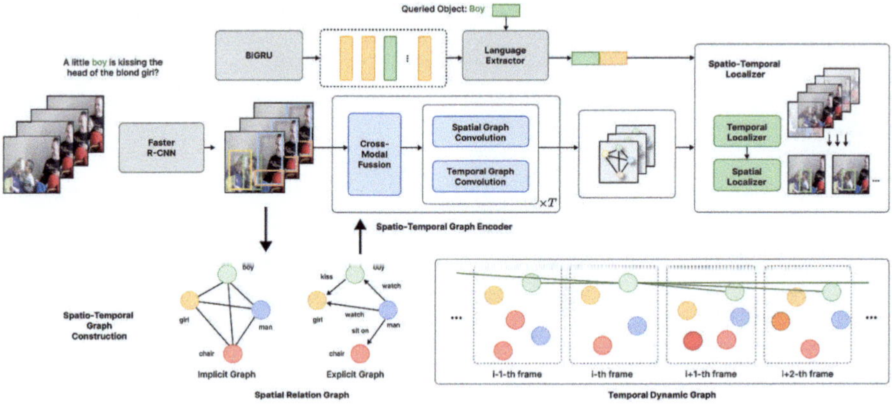

Fig. 1.51 Overview of STGRN [244] architecture. It first extracts the region features and learns the query representation. Next, it applies a spatio-temporal graph encoder to develop the multi-step cross-modal graph reasoning. After that, a spatio-temporal localizer with T spatio-temporal convolution layers directly retrieves the tubes from the region level

References

1. Ahn, D., Kim, D., Song, G., Kim, S.H., Lee, H., Kang, D., Choi, J.: Story visualization by online text augmentation with context memory. In: Proceedings of the IEEE/CVF International Conference on Computer Vision, pp. 3125–3135 (2023)
2. Ak, K.E., Lim, J.H., Tham, J.Y., Kassim, A.A.: Semantically consistent text to fashion image synthesis with an enhanced attentional generative adversarial network. Pattern Recognit. Lett. **135**, 22–29 (2020)
3. Alamri, H., Cartillier, V., Das, A., Wang, J., Cherian, A., Essa, I., Batra, D., Marks, T.K., Hori, C., Anderson, P., et al.: Audio visual scene-aware dialog. In: Proceedings of the IEEE/CVF Conference on Computer Vision and Pattern Recognition, pp. 7558–7567 (2019)
4. Alayrac, J.B., Recasens, A., Schneider, R., Arandjelović, R., Ramapuram, J., De Fauw, J., Smaira, L., Dieleman, S., Zisserman, A.: Self-supervised multimodal versatile networks. Adv. Neural Inf. Process. Syst. **33**, 25–37 (2020)
5. Alayrac, J.B., Donahue, J., Luc, P., Miech, A., Barr, I., Hasson, Y., Lenc, K., Mensch, A., Millican, K., Reynolds, M., et al.: Flamingo: a visual language model for few-shot learning. Adv. Neural Inf. Process. Syst. **35**, 23716–23736 (2022)
6. Anderson, P., He, X., Buehler, C., Teney, D., Johnson, M., Gould, S., Zhang, L.: Bottom-up and top-down attention for image captioning and visual question answering. In: Proceedings of the IEEE Conference on Computer Vision and Pattern Recognition, pp. 6077–6086 (2018)
7. Antol, S., Agrawal, A., Lu, J., Mitchell, M., Batra, D., Zitnick, C.L., Parikh, D.: Vqa: visual question answering. In: Proceedings of the IEEE International Conference on Computer Vision, pp. 2425–2433 (2015)
8. Bansal, H., Bitton, Y., Szpektor, I., Chang, K.W., Grover, A.: Videocon: robust video-language alignment via contrast captions. In: Proceedings of the IEEE/CVF Conference on Computer Vision and Pattern Recognition, pp. 13927–13937 (2024)
9. Bao, P., Zheng, Q., Mu, Y.: Dense events grounding in video. In: Proceedings of the AAAI Conference on Artificial Intelligence, vol. 35, pp. 920–928 (2021)
10. Ben-Younes, H., Cadene, R., Cord, M., Thome, N.: Mutan: multimodal tucker fusion for visual question answering. In: Proceedings of the IEEE International Conference on Computer Vision, pp. 2612–2620 (2017)

11. Blattmann, A., Rombach, R., Ling, H., Dockhorn, T., Kim, S.W., Fidler, S., Kreis, K.: Align your latents: high-resolution video synthesis with latent diffusion models. In: Proceedings of the IEEE/CVF Conference on Computer Vision and Pattern Recognition, pp. 22563–22575 (2023)
12. Brock, A., De, S., Smith, S.L., Simonyan, K.: High-performance large-scale image recognition without normalization. In: International Conference on Machine Learning, pp. 1059–1071. PMLR (2021)
13. Chen, X., Fang, H., Lin, T.Y., Vedantam, R., Gupta, S., Dollár, P., Zitnick, C.L.: Microsoft coco captions: data collection and evaluation server. arXiv preprint arXiv:1504.00325 (2015)
14. Chen, S., Jiang, W., Liu, W., Jiang, Y.G.: Learning modality interaction for temporal sentence localization and event captioning in videos. In: Computer Vision–ECCV 2020: 16th European Conference, Glasgow, August 23–28, 2020, Proceedings, Part IV 16, pp. 333–351. Springer (2020)
15. Chen, Y.C., Li, L., Yu, L., El Kholy, A., Ahmed, F., Gan, Z., Cheng, Y., Liu, J.: Uniter: universal image-text representation learning. In: European Conference on Computer Vision, pp. 104–120. Springer (2020)
16. Chen, B., Rouditchenko, A., Duarte, K., Kuehne, H., Thomas, S., Boggust, A., Panda, R., Kingsbury, B., Feris, R., Harwath, D., et al.: Multimodal clustering networks for self-supervised learning from unlabeled videos. In: Proceedings of the IEEE/CVF International Conference on Computer Vision, pp. 8012–8021 (2021)
17. Chen, F., Han, M., Zhao, H., Zhang, Q., Shi, J., Xu, S., Xu, B.: X-llm: Bootstrapping advanced large language models by treating multi-modalities as foreign languages. arXiv preprint arXiv:2305.04160 (2023)
18. Chen, H., Wang, X., Chen, H., Song, Z., Jia, J., Zhu, W.: Grounding-prompter: prompting llm with multimodal information for temporal sentence grounding in long videos. arXiv preprint arXiv:2312.17117 (2023)
19. Chen, H., Wang, X., Lan, X., Chen, H., Duan, X., Jia, J., Zhu, W.: Curriculum-listener: consistency-and complementarity-aware audio-enhanced temporal sentence grounding. In: Proceedings of the 31st ACM International Conference on Multimedia, pp. 3117–3128 (2023)
20. Chen, J., Zhu, D., Shen, X., Li, X., Liu, Z., Zhang, P., Krishnamoorthi, R., Chandra, V., Xiong, Y., Elhoseiny, M.: Minigpt-v2: large language model as a unified interface for vision-language multi-task learning. arXiv preprint arXiv:2310.09478 (2023)
21. Chen, W., Ji, Y., Wu, J., Wu, H., Xie, P., Li, J., Xia, X., Xiao, X., Lin, L.: Control-a-video: controllable text-to-video generation with diffusion models. arXiv preprint arXiv:2305.13840 (2023)
22. Chen, H., Wang, X., Chen, H., Zhang, Z., Feng, W., Huang, B., Jia, J., Zhu, W.: Verified: a video corpus moment retrieval benchmark for fine-grained video understanding. arXiv preprint arXiv:2410.08593 (2024)
23. Chen, H., Wang, X., Zhou, Y., Huang, B., Zhang, Y., Feng, W., Chen, H., Zhang, Z., Tang, S., Zhu, W.: Multi-modal generative ai: multi-modal llm, diffusion and beyond. arXiv preprint arXiv:2409.14993 (2024)
24. Cheng, B., Choudhuri, A., Misra, I., Kirillov, A., Girdhar, R., Schwing, A.G.: Mask2former for video instance segmentation. arXiv preprint arXiv:2112.10764 (2021)
25. Cho, K., Van Merriënboer, B., Gulcehre, C., Bahdanau, D., Bougares, F., Schwenk, H., Bengio, Y.: Learning phrase representations using rnn encoder-decoder for statistical machine translation. arXiv preprint arXiv:1406.1078 (2014)
26. Cho, J., Puspitasari, F.D., Zheng, S., Zheng, J., Lee, L.H., Kim, T.H., Hong, C.S., Zhang, C.: Sora as an agi world model? A complete survey on text-to-video generation. arXiv preprint arXiv:2403.05131 (2024)
27. Chollet, F.: Xception: deep learning with depthwise separable convolutions. In: Proceedings of the IEEE Conference on Computer Vision and Pattern Recognition, pp. 1251–1258 (2017)
28. Chu, W.S., Song, Y., Jaimes, A.: Video co-summarization: video summarization by visual co-occurrence. In: Proceedings of the IEEE Conference on Computer Vision and Pattern Recognition, pp. 3584–3592 (2015)

29. Chung, H.W., Hou, L., Longpre, S., Zoph, B., Tay, Y., Fedus, W., Li, Y., Wang, X., Dehghani, M., Brahma, S., et al.: Scaling instruction-finetuned language models. J. Mach. Learn. Res. **25**(70), 1–53 (2024)
30. Cui, Y., Yu, Z., Wang, C., Zhao, Z., Zhang, J., Wang, M., Yu, J.: Rosita: enhancing vision-and-language semantic alignments via cross-and intra-modal knowledge integration. In: Proceedings of the 29th ACM International Conference on Multimedia, pp. 797–806 (2021)
31. Dai, W., Li, J., Li, D., Tiong, A.M.H., Zhao, J., Wang, W., Li, B., Fung, P., Hoi, S.: Instructblip: towards general-purpose vision-language models with instruction tuning (2023)
32. Deng, C., Wu, Q., Wu, Q., Hu, F., Lyu, F., Tan, M.: Visual grounding via accumulated attention. In: Proceedings of the IEEE Conference on Computer Vision and Pattern Recognition, pp. 7746–7755 (2018)
33. Deng, J., Yang, Z., Chen, T., Zhou, W., Li, H.: Transvg: end-to-end visual grounding with transformers. In: Proceedings of the IEEE/CVF International Conference on Computer Vision, pp. 1769–1779 (2021)
34. Dhariwal, P., Nichol, A.: Diffusion models beat gans on image synthesis. Adv. Neural Inf. Process. Syst. **34**, 8780–8794 (2021)
35. Dhesikan, R., Rajmohan, V.: Sketching the future (STF): applying conditional control techniques to text-to-video models. arXiv preprint arXiv:2305.05845 (2023)
36. Ding, M., Yang, Z., Hong, W., Zheng, W., Zhou, C., Yin, D., Lin, J., Zou, X., Shao, Z., Yang, H., et al.: Cogview: mastering text-to-image generation via transformers. Adv. Neural Inf. Process. Syst. **34**, 19822–19835 (2021)
37. Donahue, J., Anne Hendricks, L., Guadarrama, S., Rohrbach, M., Venugopalan, S., Saenko, K., Darrell, T.: Long-term recurrent convolutional networks for visual recognition and description. In: Proceedings of the IEEE Conference on Computer Vision and Pattern Recognition, pp. 2625–2634 (2015)
38. Dosovitskiy, A., Beyer, L., Kolesnikov, A., Weissenborn, D., Zhai, X., Unterthiner, T., Dehghani, M., Minderer, M., Heigold, G., Gelly, S., et al.: An image is worth 16x16 words: transformers for image recognition at scale. arXiv preprint arXiv:2010.11929 (2020)
39. Duan, B., Tang, H., Wang, W., Zong, Z., Yang, G., Yan, Y.: Audio-visual event localization via recursive fusion by joint co-attention. In: Proceedings of the IEEE/CVF Winter Conference on Applications of Computer Vision, pp. 4013–4022 (2021)
40. Elasri, M., Elharrouss, O., Al-Maadeed, S., Tairi, H.: Image generation: a review. Neural Process. Lett. **54**(5), 4609–4646 (2022)
41. Escalante, H.J., Hernández, C.A., Gonzalez, J.A., López-López, A., Montes, M., Morales, E.F., Sucar, L.E., Villasenor, L., Grubinger, M.: The segmented and annotated IAPR TC-12 benchmark. Comput. Vis. Image Underst. **114**(4), 419–428 (2010)
42. Faghri, F., Fleet, D.J., Kiros, J.R., Fidler, S.: Vse++: improving visual-semantic embeddings with hard negatives. arXiv preprint arXiv:1707.05612 (2017)
43. Fan, C., Zhang, X., Zhang, S., Wang, W., Zhang, C., Huang, H.: Heterogeneous memory enhanced multimodal attention model for video question answering. In: Proceedings of the IEEE/CVF Conference on Computer Vision and Pattern Recognition, pp. 1999–2007 (2019)
44. Fang, Z., Wang, J., Hu, X., Liang, L., Gan, Z., Wang, L., Yang, Y., Liu, Z.: Injecting semantic concepts into end-to-end image captioning. In: Proceedings of the IEEE/CVF Conference on Computer Vision and Pattern Recognition, pp. 18009–18019 (2022)
45. Fang, X., Mao, K., Duan, H., Zhao, X., Li, Y., Lin, D., Chen, K.: Mmbench-video: a long-form multi-shot benchmark for holistic video understanding. arXiv preprint arXiv:2406.14515 (2024)
46. Farhadi, A., Hejrati, M., Sadeghi, M.A., Young, P., Rashtchian, C., Hockenmaier, J., Forsyth, D.: Every picture tells a story: generating sentences from images. In: Computer Vision–ECCV 2010: 11th European Conference on Computer Vision, Heraklion, September 5–11, 2010, Proceedings, Part IV 11, pp. 15–29. Springer (2010)
47. Fei, H., Wu, S., Ji, W., Zhang, H., Chua, T.S.: Empowering dynamics-aware text-to-video diffusion with large language models. arXiv preprint arXiv:2308.13812 (2023)

48. Feng, Y., Ma, L., Liu, W., Zhang, T., Luo, J.: Video re-localization. In: Proceedings of the European Conference on Computer Vision (ECCV), pp. 51–66 (2018)
49. Feng, Y., Ma, L., Liu, W., Luo, J.: Spatio-temporal video re-localization by warp lstm. In: Proceedings of the IEEE/CVF Conference on Computer Vision and Pattern Recognition, pp. 1288–1297 (2019)
50. Feng, W., Wang, X., Chen, H., Zhang, Z., Chen, H., Song, Z., Zhou, Y., Yang, Y., Wu, H., Zhu, W.: Llm4vg: large language models evaluation for video grounding (2024). https://arxiv.org/abs/2312.14206
51. Frome, A., Corrado, G.S., Shlens, J., Bengio, S., Dean, J., Ranzato, M., Mikolov, T.: Devise: a deep visual-semantic embedding model. Adv. Neural Inf. Process. Syst. **26** (2013)
52. Fu, T.J., Li, L., Gan, Z., Lin, K., Wang, W.Y., Wang, L., Liu, Z.: Violet: end-to-end video-language transformers with masked visual-token modeling. arXiv preprint arXiv:2111.12681 (2021)
53. Fu, C., Dai, Y., Luo, Y., Li, L., Ren, S., Zhang, R., Wang, Z., Zhou, C., Shen, Y., Zhang, M., et al.: Video-mme: the first-ever comprehensive evaluation benchmark of multi-modal llms in video analysis. arXiv preprint arXiv:2405.21075 (2024)
54. Fukui, A., Park, D.H., Yang, D., Rohrbach, A., Darrell, T., Rohrbach, M.: Multimodal compact bilinear pooling for visual question answering and visual grounding. arXiv preprint arXiv:1606.01847 (2016)
55. Gabeur, V., Sun, C., Alahari, K., Schmid, C.: Multi-modal transformer for video retrieval. In: Computer Vision–ECCV 2020: 16th European Conference, Glasgow, August 23–28, 2020, Proceedings, Part IV 16, pp. 214–229. Springer (2020)
56. Gan, Z., Li, L., Li, C., Wang, L., Liu, Z., Gao, J., et al.: Vision-language pre-training: basics, recent advances, and future trends. Found. Trends® Comput. Graphics Vis. **14**(3–4), 163–352 (2022)
57. Gao, H., Mao, J., Zhou, J., Huang, Z., Wang, L., Xu, W.: Are you talking to a machine? Dataset and methods for multilingual image question. Adv. Neural Inf. Process. Syst. **28**, 2296–2304 (2015)
58. Gao, J., Sun, C., Yang, Z., Nevatia, R.: TALL: temporal activity localization via language query. In: IEEE International Conference on Computer Vision, ICCV 2017, Venice, October 22–29, 2017, pp. 5277–5285 (2017)
59. Garcia, N., Vogiatzis, G.: Asymmetric spatio-temporal embeddings for large-scale image-to-video retrieval. In: 29th British Machine Vision Conference (2018)
60. Ge, S., Nah, S., Liu, G., Poon, T., Tao, A., Catanzaro, B., Jacobs, D., Huang, J.B., Liu, M.Y., Balaji, Y.: Preserve your own correlation: a noise prior for video diffusion models. In: Proceedings of the IEEE/CVF International Conference on Computer Vision, pp. 22930–22941 (2023)
61. Geng, S., Gao, P., Marks, T., Hori, C., Cherian, A.: Spatio-temporal scene graph reasoning for audio visual scene-aware dialog at DSTC8. In: DSTC8 at AAAI2020 Workshop (2020)
62. Girdhar, R., El-Nouby, A., Liu, Z., Singh, M., Alwala, K.V., Joulin, A., Misra, I.: Imagebind: one embedding space to bind them all. In: Proceedings of the IEEE/CVF Conference on Computer Vision and Pattern Recognition, pp. 15180–15190 (2023)
63. Girdhar, R., Singh, M., Brown, A., Duval, Q., Azadi, S., Rambhatla, S.S., Shah, A., Yin, X., Parikh, D., Misra, I.: Emu video: factorizing text-to-video generation by explicit image conditioning. arXiv preprint arXiv:2311.10709 (2023)
64. Gong, Y., Wang, L., Hodosh, M., Hockenmaier, J., Lazebnik, S.: Improving image-sentence embeddings using large weakly annotated photo collections. In: Computer Vision–ECCV 2014: 13th European Conference, Zurich, September 6–12, 2014, Proceedings, Part IV 13, pp. 529–545. Springer (2014)
65. Goodfellow, I., Pouget-Abadie, J., Mirza, M., Xu, B., Warde-Farley, D., Ozair, S., Courville, A., Bengio, Y.: Generative adversarial nets. Adv. Neural Inf. Process. Syst. **27**, 139–144 (2014)

66. Gupta, A., Yu, L., Sohn, K., Gu, X., Hahn, M., Fei-Fei, L., Essa, I., Jiang, L., Lezama, J.: Photorealistic video generation with diffusion models. arXiv preprint arXiv:2312.06662 (2023)
67. Hendricks, L.A., Wang, O., Shechtman, E., Sivic, J., Darrell, T., Russell, B.C.: Localizing moments in video with natural language. In: IEEE International Conference on Computer Vision, ICCV 2017, Venice, October 22–29, 2017, pp. 5804–5813 (2017)
68. Ho, J., Salimans, T.: Classifier-free diffusion guidance. arXiv preprint arXiv:2207.12598 (2022)
69. Ho, J., Salimans, T., Gritsenko, A., Chan, W., Norouzi, M., Fleet, D.J.: Video diffusion models. arXiv preprint arXiv:2204.03458 (2022)
70. Hochreiter, S., Schmidhuber, J.: Long short-term memory. Neural Comput. **9**(8), 1735–1780 (1997)
71. Hori, C., Alamri, H., Wang, J., Wichern, G., Hori, T., Cherian, A., Marks, T.K., Cartillier, V., Lopes, R.G., Das, A., et al.: End-to-end audio visual scene-aware dialog using multimodal attention-based video features. In: ICASSP 2019–2019 IEEE International Conference on Acoustics, Speech and Signal Processing (ICASSP), pp. 2352–2356. IEEE (2019)
72. Hsu, W.N., Bolte, B., Tsai, Y.H.H., Lakhotia, K., Salakhutdinov, R., Mohamed, A.: Hubert: self-supervised speech representation learning by masked prediction of hidden units. IEEE/ACM Trans. Audio Speech Lang. Process. **29**, 3451–3460 (2021)
73. Hu, R., Xu, H., Rohrbach, M., Feng, J., Saenko, K., Darrell, T.: Natural language object retrieval. In: 2016 IEEE Conference on Computer Vision and Pattern Recognition, CVPR 2016, Las Vegas, June 27–30, 2016, pp. 4555–4564 (2016)
74. Hu, Y., Luo, C., Chen, Z.: Make it move: controllable image-to-video generation with text descriptions. In: Proceedings of the IEEE/CVF Conference on Computer Vision and Pattern Recognition, pp. 18219–18228 (2022)
75. Huang, Y., Wang, W., Wang, L.: Instance-aware image and sentence matching with selective multimodal lstm. In: Proceedings of the IEEE Conference on Computer Vision and Pattern Recognition, pp. 2310–2318 (2017)
76. Huang, L., Wang, W., Chen, J., Wei, X.Y.: Attention on attention for image captioning. In: Proceedings of the IEEE/CVF International Conference on Computer Vision, pp. 4634–4643 (2019)
77. Huang, Y.H., Hsu, K.J., Jeng, S.K., Lin, Y.Y.: Weakly-supervised video re-localization with multiscale attention model. In: Proceedings of the AAAI Conference on Artificial Intelligence, vol. 34, pp. 11077–11084 (2020)
78. Huang, B., Lian, D., Luo, W., Gao, S.: Look before you leap: learning landmark features for one-stage visual grounding. In: Proceedings of the IEEE/CVF Conference on Computer Vision and Pattern Recognition, pp. 16888–16897 (2021)
79. Huang, M., Mao, Z., Wang, P., Wang, Q., Zhang, Y.: Dse-gan: dynamic semantic evolution generative adversarial network for text-to-image generation. In: Proceedings of the 30th ACM International Conference on Multimedia, pp. 4345–4354 (2022)
80. Huang, H.P., Su, Y.C., Sun, D., Jiang, L., Jia, X., Zhu, Y., Yang, M.H.: Fine-grained controllable video generation via object appearance and context. arXiv preprint arXiv:2312.02919 (2023)
81. Huang, H., Feng, Y., Shi, C., Xu, L., Yu, J., Yang, S.: Free-bloom: zero-shot text-to-video generator with llm director and ldm animator. Adv. Neural Inf. Process. Syst. **36** (2024)
82. Huang, B., Wang, X., Chen, H., Chen, H., Wu, Y., Zhu, W.: Identity-text video corpus grounding (2025)
83. Hudson, D.A., Manning, C.D.: Gqa: a new dataset for real-world visual reasoning and compositional question answering. In: Proceedings of the IEEE/CVF Conference on Computer Vision and Pattern Recognition, pp. 6700–6709 (2019)
84. Jain, Y., Nasery, A., Vineet, V., Behl, H.: Peekaboo: interactive video generation via masked-diffusion. In: Proceedings of the IEEE/CVF Conference on Computer Vision and Pattern Recognition, pp. 8079–8088 (2024)

85. Jian, Y., Gao, C., Vosoughi, S.: Bootstrapping vision-language learning with decoupled language pre-training. Adv. Neural Inf. Process. Syst. **36** (2024)
86. Jiang, C., Huang, K., He, S., Yang, X., Zhang, W., Zhang, X., Cheng, Y., Yang, L., Wang, Q., Xu, F., et al.: Learning segment similarity and alignment in large-scale content based video retrieval. In: Proceedings of the 29th ACM International Conference on Multimedia, pp. 1618–1626 (2021)
87. Jiang, H., Lin, Y., Han, D., Song, S., Huang, G.: Pseudo-q: generating pseudo language queries for visual grounding. In: Proceedings of the IEEE/CVF Conference on Computer Vision and Pattern Recognition, pp. 15513–15523 (2022)
88. Jiang, X., Xu, X., Zhang, J., Shen, F., Cao, Z., Cai, X.: Gtlr: graph-based transformer with language reconstruction for video paragraph grounding. In: 2022 IEEE International Conference on Multimedia and Expo (ICME), pp. 1–6. IEEE (2022)
89. Kamath, A., Singh, M., LeCun, Y., Synnaeve, G., Misra, I., Carion, N.: Mdetr-modulated detection for end-to-end multi-modal understanding. In: Proceedings of the IEEE/CVF International Conference on Computer Vision, pp. 1780–1790 (2021)
90. Kang, M., Zhu, J.Y., Zhang, R., Park, J., Shechtman, E., Paris, S., Park, T.: Scaling up gans for text-to-image synthesis. In: Proceedings of the IEEE/CVF Conference on Computer Vision and Pattern Recognition, pp. 10124–10134 (2023)
91. Karpathy, A., Fei-Fei, L.: Deep visual-semantic alignments for generating image descriptions. In: Proceedings of the IEEE Conference on Computer Vision and Pattern Recognition, pp. 3128–3137 (2015)
92. Karpathy, A., Joulin, A., Fei-Fei, L.F.: Deep fragment embeddings for bidirectional image sentence mapping. Adv. Neural Inf. Process. Syst. **27**, 1889–1897 (2014)
93. Kazemzadeh, S., Ordonez, V., Matten, M., Berg, T.: Referitgame: referring to objects in photographs of natural scenes. In: Proceedings of the 2014 Conference on Empirical Methods in Natural Language Processing (EMNLP), pp. 787–798 (2014)
94. Khachatryan, L., Movsisyan, A., Tadevosyan, V., Henschel, R., Wang, Z., Navasardyan, S., Shi, H.: Text2video-zero: text-to-image diffusion models are zero-shot video generators. arXiv preprint arXiv:2303.13439 (2023)
95. Kim, J.H., Lee, S.W., Kwak, D., Heo, M.O., Kim, J., Ha, J.W., Zhang, B.T.: Multimodal residual learning for visual qa. Adv. Neural Inf. Process. Syst. **29**, 361–369 (2016)
96. Kim, K.M., Heo, M.O., Choi, S.H., Zhang, B.T.: Deepstory: video story qa by deep embedded memory networks. arXiv preprint arXiv:1707.00836 (2017)
97. Kim, D., Joo, D., Kim, J.: Tivgan: text to image to video generation with step-by-step evolutionary generator. IEEE Access **8**, 153113–153122 (2020)
98. Kim, J., Yoon, S., Kim, D., Yoo, C.D.: Structured co-reference graph attention for video-grounded dialogue. In: Proceedings of the AAAI Conference on Artificial Intelligence, vol. 35, pp. 1789–1797 (2021)
99. Kiros, R., Salakhutdinov, R., Zemel, R.S.: Unifying visual-semantic embeddings with multimodal neural language models. arXiv preprint arXiv:1411.2539 (2014)
100. Krishna, R., Zhu, Y., Groth, O., Johnson, J., Hata, K., Kravitz, J., Chen, S., Kalantidis, Y., Li, L.J., Shamma, D.A., et al.: Visual genome: connecting language and vision using crowdsourced dense image annotations. Int. J. Comput. Vis. **123**, 32–73 (2017)
101. Kulkarni, G., Premraj, V., Ordonez, V., Dhar, S., Li, S., Choi, Y., Berg, A.C., Berg, T.L.: Babytalk: understanding and generating simple image descriptions. IEEE Trans. Pattern Anal. Mach. Intell. **35**(12), 2891–2903 (2013)
102. Kuznetsova, P., Ordonez, V., Berg, A., Berg, T., Choi, Y.: Collective generation of natural image descriptions. In: Proceedings of the 50th Annual Meeting of the Association for Computational Linguistics (Volume 1: Long Papers), pp. 359–368 (2012)
103. Kuznetsova, P., Ordonez, V., Berg, T.L., Choi, Y.: Treetalk: composition and compression of trees for image descriptions. Trans. Assoc. Comput. Linguist. **2**, 351–362 (2014)
104. Le, H., Chen, N.F.: Multimodal transformer with pointer network for the dstc8. In: DSTC8 at AAAI2020 Workshop (2020)

105. Le, H., Hoi, S., Sahoo, D., Chen, N.: End-to-end multimodal dialog systems with hierarchical multimodal attention on video features. In: DSTC7 at AAAI2019 Workshop (2019)
106. Le, H., Sahoo, D., Chen, N.F., Hoi, S.C.: Multimodal transformer networks for end-to-end video-grounded dialogue systems. arXiv preprint arXiv:1907.01166 (2019)
107. Lee, J.H., Kang, J.W.: Srtube: video-language pre-training with action-centric video tube features and semantic role labeling. In: Proceedings of the IEEE/CVF Conference on Computer Vision and Pattern Recognition, pp. 13689–13699 (2024)
108. Lee, K.H., Chen, X., Hua, G., Hu, H., He, X.: Stacked cross attention for image-text matching. In: Proceedings of the European Conference on Computer Vision (ECCV), pp. 201–216 (2018)
109. Lee, H., Yoon, S., Dernoncourt, F., Kim, D.S., Bui, T., Jung, K.: Dstc8-avsd: multi-modal semantic transformer network with retrieval style word generator. arXiv preprint arXiv:2004.08299 (2020)
110. Lei, J., Yu, L., Bansal, M., Berg, T.L.: Tvqa: localized, compositional video question answering. In: Empirical Methods in Natural Language Processing (2018)
111. Lei, J., Yu, L., Berg, T.L., Bansal, M.: Tvqa+: spatio-temporal grounding for video question answering. arXiv preprint arXiv:1904.11574 (2019)
112. Lei, J., Wang, L., Shen, Y., Yu, D., Berg, T.L., Bansal, M.: Mart: memory-augmented recurrent transformer for coherent video paragraph captioning. arXiv preprint arXiv:2005.05402 (2020)
113. Lei, J., Berg, T.L., Bansal, M.: Detecting moments and highlights in videos via natural language queries. Adv. Neural Inf. Process. Syst. **34**, 11846–11858 (2021)
114. Li, B., Qi, X., Lukasiewicz, T., Torr, P.: Controllable text-to-image generation. Adv. Neural Inf. Process. Syst. **32**, 2065–2075 (2019)
115. Li, X., Song, J., Gao, L., Liu, X., Huang, W., He, X., Gan, C.: Beyond rnns: positional self-attention with co-attention for video question answering. In: Proceedings of the AAAI Conference on Artificial Intelligence, vol. 33, pp. 8658–8665 (2019)
116. Li, L., Chen, Y.C., Cheng, Y., Gan, Z., Yu, L., Liu, J.: Hero: hierarchical encoder for video+language omni-representation pre-training. In: Proceedings of the 2020 Conference on Empirical Methods in Natural Language Processing (EMNLP). Association for Computational Linguistics (2020)
117. Li, J., Selvaraju, R., Gotmare, A., Joty, S., Xiong, C., Hoi, S.C.H.: Align before fuse: vision and language representation learning with momentum distillation. Adv. Neural Inf. Process. Syst. **34**, 9694–9705 (2021)
118. Li, D., Li, J., Li, H., Niebles, J.C., Hoi, S.C.: Align and prompt: video-and-language pre-training with entity prompts. In: Proceedings of the IEEE/CVF Conference on Computer Vision and Pattern Recognition, pp. 4953–4963 (2022)
119. Li, F., Zhang, H., Zhang, Y.F., Liu, S., Guo, J., Ni, L.M., Zhang, P., Zhang, L.: Vision-language intelligence: tasks, representation learning, and large models. arXiv preprint arXiv:2203.01922 (2022)
120. Li, Y., Wang, X., Xiao, J., Ji, W., Chua, T.S.: Invariant grounding for video question answering. In: Proceedings of the IEEE/CVF Conference on Computer Vision and Pattern Recognition, pp. 2928–2937 (2022)
121. Li, J., Li, D., Savarese, S., Hoi, S.: Blip-2: bootstrapping language-image pre-training with frozen image encoders and large language models. In: International Conference on Machine Learning, pp. 19730–19742. PMLR (2023)
122. Li, Y., Wang, C., Jia, J.: Llama-vid: an image is worth 2 tokens in large language models. arXiv preprint arXiv:2311.17043 (2023)
123. Li, K., He, Y., Wang, Y., Li, Y., Wang, W., Luo, P., Wang, Y., Wang, L., Qiao, Y.: Videochat: chat-centric video understanding. arXiv preprint arXiv:2305.06355 (2023)
124. Li, K., Wang, Y., He, Y., Li, Y., Wang, Y., Liu, Y., Wang, Z., Xu, J., Chen, G., Luo, P., et al.: Mvbench: a comprehensive multi-modal video understanding benchmark. In: Proceedings of the IEEE/CVF Conference on Computer Vision and Pattern Recognition, pp. 22195–22206 (2024)

125. Li, Y., Chen, X., Hu, B., Zhang, M.: Llms meet long video: advancing long video comprehension with an interactive visual adapter in llms. arXiv preprint arXiv:2402.13546 (2024)
126. Lian, L., Shi, B., Yala, A., Darrell, T., Li, B.: Llm-grounded video diffusion models. arXiv preprint arXiv:2309.17444 (2023)
127. Liang, J., Fan, Y., Zhang, K., Timofte, R., Van Gool, L., Ranjan, R.: Movideo: motion-aware video generation with diffusion models. arXiv preprint arXiv:2311.11325 (2023)
128. Liao, Y., Liu, S., Li, G., Wang, F., Chen, Y., Qian, C., Li, B.: A real-time cross-modality correlation filtering method for referring expression comprehension. In: Proceedings of the IEEE/CVF Conference on Computer Vision and Pattern Recognition, pp. 10880–10889 (2020)
129. Lin, T.Y., Maire, M., Belongie, S., Hays, J., Perona, P., Ramanan, D., Dollár, P., Zitnick, C.L.: Microsoft coco: common objects in context. In: Computer Vision–ECCV 2014: 13th European Conference, Zurich, September 6–12, 2014, Proceedings, Part V 13, pp. 740–755. Springer (2014)
130. Lin, T.Y., RoyChowdhury, A., Maji, S.: Bilinear cnns for fine-grained visual recognition. arXiv preprint arXiv:1504.07889 (2015)
131. Lin, B., Zhu, B., Ye, Y., Ning, M., Jin, P., Yuan, L.: Video-llava: learning united visual representation by alignment before projection. arXiv preprint arXiv:2311.10122 (2023)
132. Lin, H., Zala, A., Cho, J., Bansal, M.: Videodirectorgpt: consistent multi-scene video generation via llm-guided planning. arXiv preprint arXiv:2309.15091 (2023)
133. Lin, J., Yin, H., Ping, W., Molchanov, P., Shoeybi, M., Han, S.: Vila: on pre-training for visual language models. In: Proceedings of the IEEE/CVF Conference on Computer Vision and Pattern Recognition, pp. 26689–26699 (2024)
134. Liu, J., Wang, L., Yang, M.H.: Referring expression generation and comprehension via attributes. In: Proceedings of the IEEE International Conference on Computer Vision, pp. 4856–4864 (2017)
135. Liu, X., Wang, Z., Shao, J., Wang, X., Li, H.: Improving referring expression grounding with cross-modal attention-guided erasing. In: Proceedings of the IEEE/CVF Conference on Computer Vision and Pattern Recognition, pp. 1950–1959 (2019)
136. Liu, L., Li, J., Niu, L., Xu, R., Zhang, L.: Activity image-to-video retrieval by disentangling appearance and motion. In: Proceedings of the AAAI Conference on Artificial Intelligence, vol. 35, pp. 2145–2153 (2021)
137. Liu, S., Fan, H., Qian, S., Chen, Y., Ding, W., Wang, Z.: Hit: hierarchical transformer with momentum contrast for video-text retrieval. In: Proceedings of the IEEE/CVF International Conference on Computer Vision, pp. 11915–11925 (2021)
138. Liu, D., Qu, X., Zhou, P., Liu, Y.: Exploring motion and appearance information for temporal sentence grounding. In: Proceedings of the AAAI Conference on Artificial Intelligence, vol. 36, pp. 1674–1682 (2022)
139. Liu, Y., Li, S., Wu, Y., Chen, C.W., Shan, Y., Qie, X.: Umt: unified multi-modal transformers for joint video moment retrieval and highlight detection. In: Proceedings of the IEEE/CVF Conference on Computer Vision and Pattern Recognition, pp. 3042–3051 (2022)
140. Liu, Z., Ning, J., Cao, Y., Wei, Y., Zhang, Z., Lin, S., Hu, H.: Video swin transformer. In: Proceedings of the IEEE/CVF Conference on Computer Vision and Pattern Recognition, pp. 3202–3211 (2022)
141. Liu, H., Li, C., Wu, Q., Lee, Y.J.: Visual instruction tuning. arXiv preprint arXiv:2304.08485 (2023)
142. Liu, H., Li, C., Wu, Q., Lee, Y.J.: Visual instruction tuning. Adv. Neural Inf. Process. Syst. **37**, 34892–34916 (2023)
143. Liu, X., Park, D.H., Azadi, S., Zhang, G., Chopikyan, A., Hu, Y., Shi, H., Rohrbach, A., Darrell, T.: More control for free! image synthesis with semantic diffusion guidance. In: Proceedings of the IEEE/CVF Winter Conference on Applications of Computer Vision, pp. 289–299 (2023)
144. Lou, A., Lim, D., Katsman, I., Huang, L., Jiang, Q., Lim, S.N., De Sa, C.M.: Neural manifold ordinary differential equations. Adv. Neural Inf. Process. Syst. **33**, 17548–17558 (2020)

145. Lu, J., Yang, J., Batra, D., Parikh, D.: Hierarchical question-image co-attention for visual question answering. Adv. Neural Inf. Process. Syst. **29**, 289–297 (2016)
146. Lu, J., Xiong, C., Parikh, D., Socher, R.: Knowing when to look: adaptive attention via a visual sentinel for image captioning. In: Proceedings of the IEEE Conference on Computer Vision and Pattern Recognition, pp. 375–383 (2017)
147. Lu, J., Yang, J., Batra, D., Parikh, D.: Neural baby talk. In: Proceedings of the IEEE Conference on Computer Vision and Pattern Recognition, pp. 7219–7228 (2018)
148. Lu, J., Batra, D., Parikh, D., Lee, S.: Vilbert: pretraining task-agnostic visiolinguistic representations for vision-and-language tasks. Adv. Neural Inf. Process. Syst. **32**, 13–23 (2019)
149. Lu, J., Gan, R., Zhang, D., Wu, X., Wu, Z., Sun, R., Zhang, J., Zhang, P., Song, Y.: Lyrics: boosting fine-grained language-vision alignment and comprehension via semantic-aware visual objects. arXiv preprint arXiv:2312.05278 (2023)
150. Lu, Y., Zhu, L., Fan, H., Yang, Y.: Flowzero: zero-shot text-to-video synthesis with llm-driven dynamic scene syntax. arXiv preprint arXiv:2311.15813 (2023)
151. Luo, H., Ji, L., Shi, B., Huang, H., Duan, N., Li, T., Li, J., Bharti, T., Zhou, M.: Univl: a unified video and language pre-training model for multimodal understanding and generation. arXiv preprint arXiv:2002.06353 (2020)
152. Luo, R., Zhao, Z., Yang, M., Dong, J., Qiu, M., Lu, P., Wang, T., Wei, Z.: Valley: video assistant with large language model enhanced ability. arXiv preprint arXiv:2306.07207 (2023)
153. Lv, J., Huang, Y., Yan, M., Huang, J., Liu, J., Liu, Y., Wen, Y., Chen, X., Chen, S.: Gpt4motion: scripting physical motions in text-to-video generation via blender-oriented gpt planning. In: Proceedings of the IEEE/CVF Conference on Computer Vision and Pattern Recognition, pp. 1430–1440 (2024)
154. Maaz, M., Rasheed, H., Khan, S., Khan, F.S.: Video-chatgpt: towards detailed video understanding via large vision and language models. arXiv preprint arXiv:2306.05424 (2023)
155. Maharana, A., Hannan, D., Bansal, M.: Improving generation and evaluation of visual stories via semantic consistency. arXiv preprint arXiv:2105.10026 (2021)
156. Mahasseni, B., Lam, M., Todorovic, S.: Unsupervised video summarization with adversarial lstm networks. In: Proceedings of the IEEE Conference on Computer Vision and Pattern Recognition, pp. 202–211 (2017)
157. Malinowski, M., Rohrbach, M., Fritz, M.: Ask your neurons: a neural-based approach to answering questions about images. In: Proceedings of the IEEE International Conference on Computer Vision, pp. 1–9 (2015)
158. Mansimov, E., Parisotto, E., Ba, J.L., Salakhutdinov, R.: Generating images from captions with attention. arXiv preprint arXiv:1511.02793 (2015)
159. Mao, J., Xu, W., Yang, Y., Wang, J., Huang, Z., Yuille, A.: Deep captioning with multimodal recurrent neural networks (m-rnn). arXiv preprint arXiv:1412.6632 (2014)
160. Mao, J., Huang, J., Toshev, A., Camburu, O., Yuille, A.L., Murphy, K.: Generation and comprehension of unambiguous object descriptions. In: Proceedings of the IEEE Conference on Computer Vision and Pattern Recognition, pp. 11–20 (2016)
161. Miech, A., Alayrac, J.B., Smaira, L., Laptev, I., Sivic, J., Zisserman, A.: End-to-end learning of visual representations from uncurated instructional videos. In: Proceedings of the IEEE/CVF Conference on Computer Vision and Pattern Recognition, pp. 9879–9889 (2020)
162. Mirza, M.: Conditional generative adversarial nets. arXiv preprint arXiv:1411.1784 (2014)
163. Mo, S., Wang, H., Li, H., Tang, X.: Unified video-language pre-training with synchronized audio. arXiv preprint arXiv:2405.07202 (2024)
164. Nagaraja, V.K., Morariu, V.I., Davis, L.S.: Modeling context between objects for referring expression understanding. In: Computer Vision–ECCV 2016: 14th European Conference, Amsterdam, October 11–14, 2016, Proceedings, Part IV 14, pp. 792–807. Springer (2016)
165. Nam, H., Ha, J.W., Kim, J.: Dual attention networks for multimodal reasoning and matching. In: Proceedings of the IEEE Conference on Computer Vision and Pattern Recognition, pp. 299–307 (2017)

166. Nguyen, D.T., Sharma, S., Schulz, H., Asri, L.E.: From film to video: multi-turn question answering with multi-modal context. arXiv preprint arXiv:1812.07023 (2018)
167. Nichol, A., Dhariwal, P., Ramesh, A., Shyam, P., Mishkin, P., McGrew, B., Sutskever, I., Chen, M.: Glide: towards photorealistic image generation and editing with text-guided diffusion models. arXiv preprint arXiv:2112.10741 (2021)
168. OpenAI: Gpt-4 technical report (2023)
169. Pan, Y., Mei, T., Yao, T., Li, H., Rui, Y.: Jointly modeling embedding and translation to bridge video and language. In: Proceedings of the IEEE Conference on Computer Vision and Pattern Recognition, pp. 4594–4602 (2016)
170. Pan, Y., Yao, T., Li, H., Mei, T.: Video captioning with transferred semantic attributes. In: Proceedings of the IEEE Conference on Computer Vision and Pattern Recognition, pp. 6504–6512 (2017)
171. Pan, J., Goyal, Y., Lee, S.: Question-conditioned counterfactual image generation for vqa. arXiv preprint arXiv:1911.06352 (2019)
172. Papalampidi, P., Koppula, S., Pathak, S., Chiu, J., Heyward, J., Patraucean, V., Shen, J., Miech, A., Zisserman, A., Nematzdeh, A.: A simple recipe for contrastively pre-training video-first encoders beyond 16 frames. In: Proceedings of the IEEE/CVF Conference on Computer Vision and Pattern Recognition, pp. 14386–14397 (2024)
173. Plummer, B.A., Wang, L., Cervantes, C.M., Caicedo, J.C., Hockenmaier, J., Lazebnik, S.: Flickr30k entities: collecting region-to-phrase correspondences for richer image-to-sentence models. In: Proceedings of the IEEE International Conference on Computer Vision, pp. 2641–2649 (2015)
174. Qiao, T., Zhang, J., Xu, D., Tao, D.: Learn, imagine and create: text-to-image generation from prior knowledge. Adv. Neural Inf. Process. Syst. **32**, 887–897 (2019)
175. Qiu, H., Xia, M., Zhang, Y., He, Y., Wang, X., Shan, Y., Liu, Z.: Freenoise: tuning-free longer video diffusion via noise rescheduling. arXiv preprint arXiv:2310.15169 (2023)
176. Ramesh, A., Pavlov, M., Goh, G., Gray, S., Voss, C., Radford, A., Chen, M., Sutskever, I.: Zero-shot text-to-image generation. In: International Conference on Machine Learning, pp. 8821–8831. PMLR (2021)
177. Ramesh, A., Dhariwal, P., Nichol, A., Chu, C., Chen, M.: Hierarchical text-conditional image generation with clip latents. arXiv preprint arXiv:2204.06125 (2022)
178. Ren, M., Kiros, R., Zemel, R.: Exploring models and data for image question answering. Adv. Neural Inf. Process. Syst. **28**, 2953–2961 (2015)
179. Ren, S., He, K., Girshick, R., Sun, J.: Faster r-cnn: towards real-time object detection with region proposal networks. Adv. Neural Inf. Process. Syst. **28**, 91–99 (2015)
180. Rombach, R., Blattmann, A., Lorenz, D., Esser, P., Ommer, B.: High-resolution image synthesis with latent diffusion models. In: Proceedings of the IEEE/CVF Conference on Computer Vision and Pattern Recognition, pp. 10684–10695 (2022)
181. Sadhu, A., Chen, K., Nevatia, R.: Zero-shot grounding of objects from natural language queries. In: Proceedings of the IEEE/CVF International Conference on Computer Vision, pp. 4694–4703 (2019)
182. Saharia, C., Chan, W., Saxena, S., Li, L., Whang, J., Denton, E.L., Ghasemipour, K., Gontijo Lopes, R., Karagol Ayan, B., Salimans, T., et al.: Photorealistic text-to-image diffusion models with deep language understanding. Adv. Neural Inf. Process. Syst. **35**, 36479–36494 (2022)
183. Sharghi, A., Laurel, J.S., Gong, B.: Query-focused video summarization: dataset, evaluation, and a memory network based approach. In: Proceedings of the IEEE Conference on Computer Vision and Pattern Recognition, pp. 4788–4797 (2017)
184. Sharma, S., Suhubdy, D., Michalski, V., Kahou, S.E., Bengio, Y.: Chatpainter: improving text to image generation using dialogue. arXiv preprint arXiv:1802.08216 (2018)
185. Shen, R., Inoue, N., Shinoda, K.: Text-guided object detector for multi-modal video question answering. In: Proceedings of the IEEE/CVF Winter Conference on Applications of Computer Vision, pp. 1032–1042 (2023)

186. Shi, F., Wang, L., Huang, W.: End-to-end dense video grounding via parallel regression. arXiv preprint arXiv:2109.11265 (2021)
187. Shih, K.J., Singh, S., Hoiem, D.: Where to look: focus regions for visual question answering. In: Proceedings of the IEEE Conference on Computer Vision and Pattern Recognition, pp. 4613–4621 (2016)
188. Simonyan, K., Zisserman, A.: Very deep convolutional networks for large-scale image recognition. arXiv preprint arXiv:1409.1556 (2014)
189. Singer, U., Polyak, A., Hayes, T., Yin, X., An, J., Zhang, S., Hu, Q., Yang, H., Ashual, O., Gafni, O., et al.: Make-a-video: text-to-video generation without text-video data. arXiv preprint arXiv:2209.14792 (2022)
190. Socher, R., Karpathy, A., Le, Q.V., Manning, C.D., Ng, A.Y.: Grounded compositional semantics for finding and describing images with sentences. Trans. Assoc. Comput. Linguist. **2**, 207–218 (2014)
191. Sohl-Dickstein, J., Weiss, E., Maheswaranathan, N., Ganguli, S.: Deep unsupervised learning using nonequilibrium thermodynamics. In: International Conference on Machine Learning, pp. 2256–2265. PMLR (2015)
192. Song, Y., Sohl-Dickstein, J., Kingma, D.P., Kumar, A., Ermon, S., Poole, B.: Score-based generative modeling through stochastic differential equations. arXiv preprint arXiv:2011.13456 (2020)
193. Song, E., Chai, W., Wang, G., Zhang, Y., Zhou, H., Wu, F., Guo, X., Ye, T., Lu, Y., Hwang, J.N., et al.: Moviechat: from dense token to sparse memory for long video understanding. arXiv preprint arXiv:2307.16449 (2023)
194. Sun, C., Myers, A., Vondrick, C., Murphy, K., Schmid, C.: Videobert: a joint model for video and language representation learning. In: Proceedings of the IEEE/CVF International Conference on Computer Vision, pp. 7464–7473 (2019)
195. Sutskever, I., Vinyals, O., Le, Q.V.: Sequence to sequence learning with neural networks. Adv. Neural Inf. Process. Syst. **27**, 3104–3112 (2014)
196. Szegedy, C., Liu, W., Jia, Y., Sermanet, P., Reed, S., Anguelov, D., Erhan, D., Vanhoucke, V., Rabinovich, A.: Going deeper with convolutions. In: Proceedings of the IEEE Conference on Computer Vision and Pattern Recognition, pp. 1–9 (2015)
197. Tan, H., Bansal, M.: Lxmert: learning cross-modality encoder representations from transformers. arXiv preprint arXiv:1908.07490 (2019)
198. Tang, W., Li, L., Liu, X., Jin, L., Tang, J., Li, Z.: Context disentangling and prototype inheriting for robust visual grounding. In: IEEE Transactions on Pattern Analysis and Machine Intelligence (2023)
199. Tao, M., Bao, B.K., Tang, H., Xu, C.: Galip: generative adversarial clips for text-to-image synthesis. In: Proceedings of the IEEE/CVF Conference on Computer Vision and Pattern Recognition, pp. 14214–14223 (2023)
200. Tapaswi, M., Zhu, Y., Stiefelhagen, R., Torralba, A., Urtasun, R., Fidler, S.: Movieqa: understanding stories in movies through question-answering. In: Proceedings of the IEEE Conference on Computer Vision and Pattern Recognition, pp. 4631–4640 (2016)
201. Tay, Y., Dehghani, M., Tran, V.Q., Garcia, X., Wei, J., Wang, X., Chung, H.W., Shakeri, S., Bahri, D., Schuster, T., et al.: Ul2: unifying language learning paradigms. arXiv preprint arXiv:2205.05131 (2022)
202. THUNLP ModelBest, Z.: Viscpm: Chinese-English bilingual multi-modal large model series (2023). https://github.com/OpenBMB/VisCPM
203. Tian, Y., Shi, J., Li, B., Duan, Z., Xu, C.: Audio-visual event localization in unconstrained videos. In: Proceedings of the European Conference on Computer Vision (ECCV), pp. 247–263 (2018)
204. Touvron, H., Lavril, T., Izacard, G., Martinet, X., Lachaux, M.A., Lacroix, T., Rozière, B., Goyal, N., Hambro, E., Azhar, F., et al.: Llama: open and efficient foundation language models. arXiv preprint arXiv:2302.13971 (2023)
205. Vinyals, O., Toshev, A., Bengio, S., Erhan, D.: Show and tell: a neural image caption generator. In: Proceedings of the IEEE Conference on Computer Vision and Pattern Recognition, pp. 3156–3164 (2015)

206. Wang, P., Wu, Q., Cao, J., Shen, C., Gao, L., Hengel, A.v.d.: Neighbourhood watch: referring expression comprehension via language-guided graph attention networks. In: Proceedings of the IEEE/CVF Conference on Computer Vision and Pattern Recognition, pp. 1960–1968 (2019)
207. Wang, L., Huang, J., Li, Y., Xu, K., Yang, Z., Yu, D.: Improving weakly supervised visual grounding by contrastive knowledge distillation. In: Proceedings of the IEEE/CVF Conference on Computer Vision and Pattern Recognition, pp. 14090–14100 (2021)
208. Wang, J., Ge, Y., Cai, G., Yan, R., Lin, X., Shan, Y., Qie, X., Shou, M.Z.: Object-aware video-language pre-training for retrieval. In: Proceedings of the IEEE/CVF Conference on Computer Vision and Pattern Recognition, pp. 3313–3322 (2022)
209. Wang, J., Yang, Z., Hu, X., Li, L., Lin, K., Gan, Z., Liu, Z., Liu, C., Wang, L.: Git: a generative image-to-text transformer for vision and language. arXiv preprint arXiv:2205.14100 (2022)
210. Wang, R., Chen, D., Wu, Z., Chen, Y., Dai, X., Liu, M., Jiang, Y.G., Zhou, L., Yuan, L.: Bevt: Bert pretraining of video transformers. In: Proceedings of the IEEE/CVF Conference on Computer Vision and Pattern Recognition, pp. 14733–14743 (2022)
211. Wang, Y., He, Y., Li, Y., Li, K., Yu, J., Ma, X., Chen, X., Wang, Y., Luo, P., Liu, Z., et al.: Internvid: a large-scale video-text dataset for multimodal understanding and generation. arXiv preprint arXiv:2307.06942 (2023)
212. Wang, Y., Wang, Y., Wu, P., Liang, J., Zhao, D., Zheng, Z.: Lstp: language-guided spatial-temporal prompt learning for long-form video-text understanding. arXiv preprint arXiv:2402.16050 (2024)
213. Wu, Y., Zhu, L., Yan, Y., Yang, Y.: Dual attention matching for audio-visual event localization. In: Proceedings of the IEEE/CVF International Conference on Computer Vision, pp. 6292–6300 (2019)
214. Wu, C., Liang, J., Ji, L., Yang, F., Fang, Y., Jiang, D., Duan, N.: Nüwa: visual synthesis pre-training for neural visual world creation. In: European Conference on Computer Vision, pp. 720–736. Springer (2022)
215. Wu, W., Luo, H., Fang, B., Wang, J., Ouyang, W.: Cap4video: What can auxiliary captions do for text-video retrieval? In: Proceedings of the IEEE/CVF Conference on Computer Vision and Pattern Recognition, pp. 10704–10713 (2023)
216. Xu, K., Ba, J., Kiros, R., Cho, K., Courville, A., Salakhudinov, R., Zemel, R., Bengio, Y.: Show, attend and tell: neural image caption generation with visual attention. In: International Conference on Machine Learning, pp. 2048–2057. PMLR (2015)
217. Xu, J., Mei, T., Yao, T., Rui, Y.: Msr-vtt: a large video description dataset for bridging video and language. In: Proceedings of the IEEE Conference on Computer Vision and Pattern Recognition, pp. 5288–5296 (2016)
218. Xu, D., Zhao, Z., Xiao, J., Wu, F., Zhang, H., He, X., Zhuang, Y.: Video question answering via gradually refined attention over appearance and motion. In: Proceedings of the 25th ACM International Conference on Multimedia, pp. 1645–1653 (2017)
219. Xu, T., Zhang, P., Huang, Q., Zhang, H., Gan, Z., Huang, X., He, X.: Attngan: fine-grained text to image generation with attentional generative adversarial networks. In: Proceedings of the IEEE Conference on Computer Vision and Pattern Recognition, pp. 1316–1324 (2018)
220. Xu, H., Zeng, R., Wu, Q., Tan, M., Gan, C.: Cross-modal relation-aware networks for audio-visual event localization. In: Proceedings of the 28th ACM International Conference on Multimedia, pp. 3893–3901 (2020)
221. Xu, R., Niu, L., Zhang, J., Zhang, L.: A proposal-based approach for activity image-to-video retrieval. In: Proceedings of the AAAI Conference on Artificial Intelligence, vol. 34, pp. 12524–12531 (2020)
222. Xu, J., Lan, C., Xie, W., Chen, X., Lu, Y.: Retrieval-based video language model for efficient long video question answering. arXiv preprint arXiv:2312.04931 (2023)
223. Xuan, H., Zhang, Z., Chen, S., Yang, J., Yan, Y.: Cross-modal attention network for temporal inconsistent audio-visual event localization. In: Proceedings of the AAAI Conference on Artificial Intelligence, vol. 34, pp. 279–286 (2020)

224. Yamaguchi, M., Saito, K., Ushiku, Y., Harada, T.: Spatio-temporal person retrieval via natural language queries. In: Proceedings of the IEEE International Conference on Computer Vision, pp. 1453–1462 (2017)
225. Yang, Z., He, X., Gao, J., Deng, L., Smola, A.: Stacked attention networks for image question answering. In: Proceedings of the IEEE Conference on Computer Vision and Pattern Recognition, pp. 21–29 (2016)
226. Yang, S., Li, G., Yu, Y.: Graph-structured referring expression reasoning in the wild. In: Proceedings of the IEEE/CVF Conference on Computer Vision and Pattern Recognition, pp. 9952–9961 (2020)
227. Yang, A., Miech, A., Sivic, J., Laptev, I., Schmid, C.: Just ask: learning to answer questions from millions of narrated videos. In: Proceedings of the IEEE/CVF International Conference on Computer Vision, pp. 1686–1697 (2021)
228. Yang, A., Miech, A., Sivic, J., Laptev, I., Schmid, C.: Learning to answer visual questions from web videos. arXiv preprint arXiv:2205.05019 (2022)
229. Yang, A., Miech, A., Sivic, J., Laptev, I., Schmid, C.: Tubedetr: spatio-temporal video grounding with transformers. In: Proceedings of the IEEE/CVF Conference on Computer Vision and Pattern Recognition, pp. 16442–16453 (2022)
230. Yang, A., Miech, A., Sivic, J., Laptev, I., Schmid, C.: Zero-shot video question answering via frozen bidirectional language models. Adv. Neural Inf. Process. Syst. **35**, 124–141 (2022)
231. Yao, T., Pan, Y., Li, Y., Qiu, Z., Mei, T.: Boosting image captioning with attributes. In: Proceedings of the IEEE International Conference on Computer Vision, pp. 4894–4902 (2017)
232. Yao, T., Pan, Y., Li, Y., Mei, T.: Exploring visual relationship for image captioning. In: Proceedings of the European Conference on Computer Vision (ECCV), pp. 684–699 (2018)
233. Yin, G., Liu, B., Sheng, L., Yu, N., Wang, X., Shao, J.: Semantics disentangling for text-to-image generation. In: Proceedings of the IEEE/CVF Conference on Computer Vision and Pattern Recognition, pp. 2327–2336 (2019)
234. Yu, L., Lin, Z., Shen, X., Yang, J., Lu, X., Bansal, M., Berg, T.L.: Mattnet: modular attention network for referring expression comprehension. In: Proceedings of the IEEE Conference on Computer Vision and Pattern Recognition, pp. 1307–1315 (2018)
235. Yu, Z., Yu, J., Cui, Y., Tao, D., Tian, Q.: Deep modular co-attention networks for visual question answering. In: Proceedings of the IEEE/CVF Conference on Computer Vision and Pattern Recognition, pp. 6281–6290 (2019)
236. Yu, J., Wang, Z., Vasudevan, V., Yeung, L., Seyedhosseini, M., Wu, Y.: Coca: contrastive captioners are image-text foundation models. arXiv preprint arXiv:2205.01917 (2022)
237. Yu, J., Xu, Y., Koh, J.Y., Luong, T., Baid, G., Wang, Z., Vasudevan, V., Ku, A., Yang, Y., Ayan, B.K., et al.: Scaling autoregressive models for content-rich text-to-image generation. arXiv preprint arXiv:2206.10789 (2022)
238. Yuan, Y., Mei, T., Cui, P., Zhu, W.: Video summarization by learning deep side semantic embedding. IEEE Trans. Circuits Syst. Video Technol. **29**(1), 226–237 (2017)
239. Zellers, R., Lu, X., Hessel, J., Yu, Y., Park, J.S., Cao, J., Farhadi, A., Choi, Y.: Merlot: multimodal neural script knowledge models. Adv. Neural Inf. Process. Syst. **34**, 23634–23651 (2021)
240. Zellers, R., Lu, J., Lu, X., Yu, Y., Zhao, Y., Salehi, M., Kusupati, A., Hessel, J., Farhadi, A., Choi, Y.: Merlot reserve: neural script knowledge through vision and language and sound. In: Proceedings of the IEEE/CVF Conference on Computer Vision and Pattern Recognition, pp. 16375–16387 (2022)
241. Zeng, A., Liu, X., Du, Z., Wang, Z., Lai, H., Ding, M., Yang, Z., Xu, Y., Zheng, W., Xia, X., et al.: Glm-130b: an open bilingual pre-trained model. arXiv preprint arXiv:2210.02414 (2022)
242. Zhai, M., Chen, L., Tung, F., He, J., Nawhal, M., Mori, G.: Lifelong gan: continual learning for conditional image generation. In: Proceedings of the IEEE/CVF International Conference on Computer Vision, pp. 2759–2768 (2019)

243. Zhang, Z., Zhao, Z., Lin, Z., Song, J., Cai, D.: Localizing unseen activities in video via image query. arXiv preprint arXiv:1906.12165 (2019)
244. Zhang, Z., Zhao, Z., Zhao, Y., Wang, Q., Liu, H., Gao, L.: Where does it exist: spatio-temporal video grounding for multi-form sentences. In: Proceedings of the IEEE/CVF Conference on Computer Vision and Pattern Recognition, pp. 10668–10677 (2020)
245. Zhang, Y., Li, X., Liu, C., Shuai, B., Zhu, Y., Brattoli, B., Chen, H., Marsic, I., Tighe, J.: Vidtr: video transformer without convolutions. In: 2021 IEEE/CVF International Conference on Computer Vision, ICCV 2021, Montreal, October 10–17, 2021, pp. 13557–13567. IEEE (2021). https://doi.org/10.1109/ICCV48922.2021.01332.
246. Zhang, H., Koh, J.Y., Baldridge, J., Lee, H., Yang, Y.: Cross-modal contrastive learning for text-to-image generation. In: Proceedings of the IEEE/CVF Conference on Computer Vision and Pattern Recognition, pp. 833–842 (2021)
247. Zhang, C., Lu, T., Islam, M.M., Wang, Z., Yu, S., Bansal, M., Bertasius, G.: A simple llm framework for long-range video question-answering. In: Proceedings of the 2024 Conference on Empirical Methods in Natural Language Processing (EMNLP), pp. 21715–21737 (2024)
248. Zhang, C., Zhang, C., Zhang, M., Kweon, I.S.: Text-to-image diffusion model in generative ai: a survey. arXiv preprint arXiv:2303.07909 (2023)
249. Zhang, H., Li, X., Bing, L.: Video-llama: an instruction-tuned audio-visual language model for video understanding. arXiv preprint arXiv:2306.02858 (2023)
250. Zhang, H., Sun, A., Jing, W., Zhou, J.T.: Temporal sentence grounding in videos: a survey and future directions. In: IEEE Transactions on Pattern Analysis and Machine Intelligence (2023)
251. Zhang, Y., Han, S., Zhang, Z., Wang, J., Bi, H.: Cf-gan: cross-domain feature fusion generative adversarial network for text-to-image synthesis. Vis. Comput. **39**(4), 1283–1293 (2023)
252. Zhang, D., Yu, Y., Li, C., Dong, J., Su, D., Chu, C., Yu, D.: Mm-llms: recent advances in multimodal large language models. arXiv preprint arXiv:2401.13601 (2024)
253. Zhang, Y., Zhao, Z., Chen, Z., Feng, Z., Ding, Z., Sun, Y.: Rankclip: ranking-consistent language-image pretraining. arXiv preprint arXiv:2404.09387 (2024)
254. Zhao, B., Li, X., Lu, X.: Hsa-rnn: hierarchical structure-adaptive rnn for video summarization. In: Proceedings of the IEEE Conference on Computer Vision and Pattern Recognition, pp. 7405–7414 (2018)
255. Zhao, Y., Misra, I., Krähenbühl, P., Girdhar, R.: Learning video representations from large language models. In: Proceedings of the IEEE/CVF Conference on Computer Vision and Pattern Recognition, pp. 6586–6597 (2023)
256. Zheng, K., Zhang, Y., Wu, W., Lu, F., Ma, S., Jin, X., Chen, W., Shen, Y.: Dreamlip: language-image pre-training with long captions. arXiv preprint arXiv:2403.17007 (2024)
257. Zhou, L., Palangi, H., Zhang, L., Hu, H., Corso, J., Gao, J.: Unified vision-language pre-training for image captioning and vqa. In: Proceedings of the AAAI Conference on Artificial intelligence, vol. 34, pp. 13041–13049 (2020)
258. Zhu, L., Yang, Y.: Actbert: learning global-local video-text representations. In: Proceedings of the IEEE/CVF Conference on Computer Vision and Pattern Recognition, pp. 8746–8755 (2020)
259. Zhu, B., Lin, B., Ning, M., Yan, Y., Cui, J., Wang, H., Pang, Y., Jiang, W., Zhang, J., Li, Z., et al.: Languagebind: extending video-language pretraining to n-modality by language-based semantic alignment. arXiv preprint arXiv:2310.01852 (2023)
260. Zhu, D., Chen, J., Shen, X., Li, X., Elhoseiny, M.: Minigpt-4: enhancing vision-language understanding with advanced large language models. arXiv preprint arXiv:2304.10592 (2023)
261. Zitnick, C.L., Dollár, P.: Edge boxes: locating object proposals from edges. In: Computer Vision–ECCV 2014: 13th European Conference, Zurich, September 6-12, 2014, Proceedings, Part V 13, pp. 391–405. Springer (2014)

Chapter 2
Basics and Advances of Video Grounding

Abstract In this chapter, we give a comprehensive overview of *Video Grounding*, which (i) summarizes the taxonomy of existing methods, (ii) provides a detailed description of the evaluation protocols (i.e., datasets and metrics) to be used in *Video Grounding*, and (iii) discusses thoroughly some advances in *Video Grounding*.

2.1 Video Grounding Models

We establish the taxonomy of existing approaches based on their characteristics (c.f., Fig. 2.1). Early works adopt a two-stage architecture (c.f., Fig. 2.2a). To be more specific, they first scan the whole video and pre-cut various candidate segments (i.e., proposals or moments) using a sliding window strategy or proposal generation network, and then rank the candidates according to the ranking scores produced by the cross-modal matching module. However, the *scan-and-localize* pipeline is computationally intensive due to excessive redundant calculations for overlapping candidate segments. Additionally, individual pairwise segment-query matching may overlook crucial contextual information within the video.

Considering the above concerns, some researchers start to use single-stage methods to solve *Video Grounding* without pre-cutting candidate moments (c.f., Fig. 2.2b). Instead, multi-scale candidate moments ended at each time step are maintained by LSTM sequentially or convolutional neural networks hierarchically, which are named anchor-based methods. Some other single-stage methods predict the probabilities for each video unit (i.e., frame-level or clip-level) being the start and end point of the target segment, or straightforwardly regress the target start and end coordinates based on the multi-modal feature of the providing video and sentence query. These methods do not depend on any candidate proposal generation process, and are named anchor-free methods.

Besides, it is worth noting that some works resort to deep reinforcement learning techniques to address *Video Grounding*, taking the sentence localization problem as a sequential decision process, which are also anchor-free. To reduce the intensive labor required for annotating the boundaries of ground truth moments, weakly supervised methods with only video-level annotated descriptions have also emerged,

X. Wang et al., *Video Grounding and Its Generalization*,
https://doi.org/10.1007/978-3-031-94837-4_2

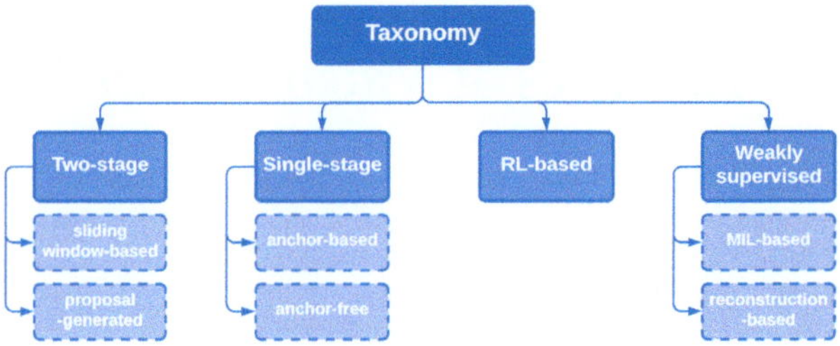

Fig. 2.1 The taxonomy of existing approaches, grouped into early two-stage methods, typical single-stage methods, reinforcement learning (RL)-based methods, and weakly supervised methods. According to the ways of proposal generation, the two-stage methods can be subsequently divided into sliding window-based and proposal-generated ones. Meanwhile, the single-stage methods can be divided into anchor-based and anchor-free ones, which depends on whether or not anchors (candidate moments) are produced for ranking. The weakly supervised methods could also be further grouped into MIL-based and reconstruction-based

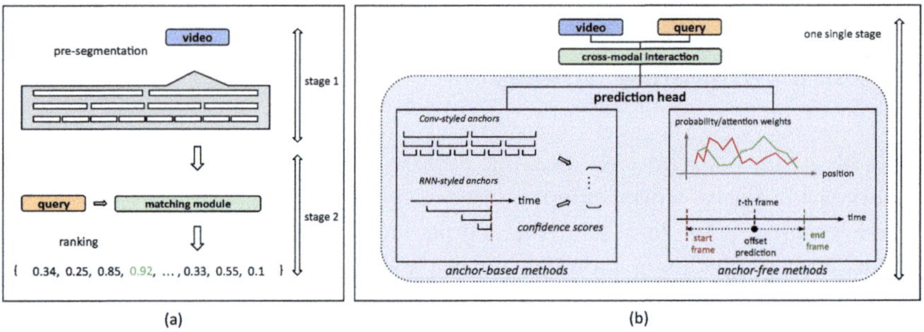

Fig. 2.2 Two-stage methods vs. Single-stage methods. (**a**) At stage 1, the entire video is pre-segmented into multi-scale candidate moments. At stage 2, the matching module takes query-moment pairs as inputs and outputs matching scores for ranking. (**b**) Single-stage methods can be either anchor-based or anchor-free. Anchor-based methods use different types of anchors (e.g., Conv-styled, RNN-styled) as candidate moments, while anchor-free methods use prediction head to directly obtain moment boundaries (e.g., generate probability/attention weights for each position, or predict the offsets from a certain frame to the start/end ground truth boundaries)

which can be either MIL-based or reconstruction-based. In the following, we will present all the approaches and perform a deep analysis of the characteristics for each type.

2.1.1 *Fully Supervised Methods*

First of all, we discuss *Video Grounding* approaches designed under the supervised learning paradigm with full labels.

2.1.1.1 Two-Stage Methods

For a two-stage method, the pre-segmenting of proposal candidates is conducted separately with the model computation. It takes the pre-segmented candidates and the sentence query as inputs of a cross-modal matching module for target segment localization. The two-stage methods can be grouped into two categories based on different ways to generate proposals.

Sliding-Window Based

Early methods adopt a multi-scale sliding window sampling strategy for the generation of candidate proposals. There are two pioneering works, MCN [45] and CTRL [36], to define the *Video Grounding* task and construct benchmark datasets. First, Hendricks et al. [45] proposes MCN, which uses a sliding window mechanism to sample all the candidate moments (i.e. segments) before projecting the query representation and video moment representation into a shared embedding space. The ℓ_2 distance between the sentence query and the corresponding target video moment in this space is minimized to supervise the model training (c.f., Fig. 2.3b). Specifically, MCN encourages the sentence query to be closer to the target moment than negative moments in a shared embedding space. Since the negative moments either come from other segments within the same video (intra-video) or from different videos (inter-video), MCN creates two ranking loss functions that are similar but distinct:

$$
\begin{aligned}
\mathcal{L}_i^{intra}(\theta) &= \sum_{n \in \Gamma \setminus \tau^i} \mathcal{L}^R(D_\theta(s^i, v^i, \tau^i), D_\theta(s^i, v^i, n)), \\
\mathcal{L}_i^{inter}(\theta) &= \sum_{j \neq i} \mathcal{L}^R(D_\theta(s^i, v^i, \tau^i), D_\theta(s^i, v^j, \tau^i)),
\end{aligned}
\tag{2.1}
$$

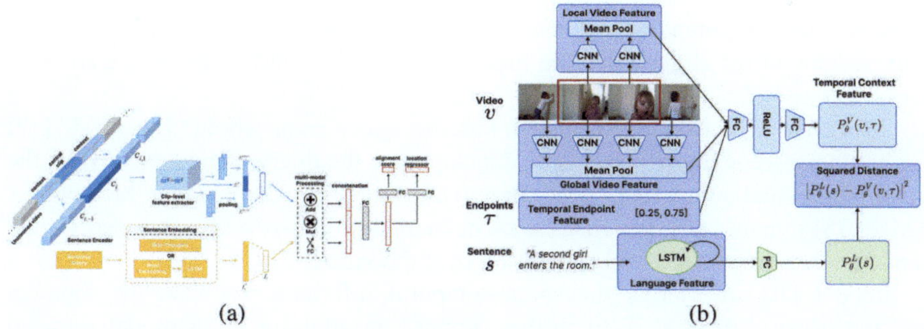

Fig. 2.3 The (**a**) CTRL and (**b**) MCN frameworks, two pioneer works that firstly present *Video Grounding* task. CTRL uses joint language-segment representations to get the final alignment scores and refines the temporal boundaries by location regressor, while MCN tries to minimize the ℓ_2 distance between the language and video representations in a common space

where $\mathcal{L}^R(x, y) = \max(0, x - y + b)$, b is a margin. As for training sample i, the intra-video ranking loss encourages sentence i to be closer to the target moment at the location τ^i than the negative moments from other possible locations within the same video, while the inter-video ranking loss encourages sentence i to be closer to the target one at location τ^i than the negative ones from other videos of the same location τ^i. The intra-video ranking loss is able to differentiate between subtle differences within a video while the inter-video ranking loss can differentiate between broad semantic concepts.

At the same time, Gao et al. [36] propose CTRL, which is the first to adapt R-CNN [40] methodology from object detection to the *Video Grounding* domain. Particularly, CTRL also leverages a sliding window to obtain candidate segments of various lengths, and as shown in Fig. 2.3a, it exploits a multi-modal processing module to fuse the candidate segment representation with the sentence representation by three operators (i.e., add, multiply, and fully-connected layer). Then, CTRL feeds the fused representation into another fully-connected layer to predict the alignment score and location offsets between the candidate segment and the target segment. CTRL designs a multi-task loss function to train the model, including visual-semantic alignment loss and location regression loss:

$$\mathcal{L}_{aln} = \frac{1}{N}\sum_{i=0}^{N}[\alpha_c \log(1 + \exp(-cs_{i,i})) + \sum_{j=0, j\neq i}^{N} \alpha_w \log(1 + \exp(cs_{i,j}))] , \tag{2.2}$$

$$\mathcal{L}_{reg} = \frac{1}{N}\sum_{i=0}^{N}[R(t^*_{x,i} - t_{x,i}) + R(t^*_{y,i} - t_{y,i})] , \tag{2.3}$$

where $\mathcal{L}_{aln}$ is the visual-semantic alignment loss considering both aligned (video segment, query) pairs and misaligned pairs. $cs_{i,j}$ measures the alignment score between video segment c_i and sentence s_j. The location regression loss $\mathcal{L}_{reg}$ is only accounted for aligned pairs to predict the correct coordinates. R is a smooth-L1 function.

Compared to above CTRL that treats the query as a whole, Liu et al. [77] further make some improvements by decomposing the query and adaptively get the important textual components according to the temporal video context. Meanwhile, TMN [75] dynamically generates a modular neural network layout based on the semantic structure of the query to reason over the video.

Since CTRL overlooks the spatio-temporal information inside the moment and the query, Liu et al. [76] further propose an attentive cross-modal retrieval network (ACRN). With a memory attention network guided by the sentence query, ACRN adaptively assigns weights to the contextual moment representations for memorization, augmenting the moment representation. SLTA [55] also devises a spatial and language-temporal attention model to adaptively identify the relevant objects and interactions based on the query information. Considering that the inherent spatio-temporal structure of videos can not be fully captured by one-dimensional vectors in CTRL, Song et al. [115] propose to employ voxel- and channel-wise attention over the visual 3D feature maps to improve visual features and cross-modal correlation.

Wu and Han [139] propose a multi-modal circulant fusion (MCF) in contrast to the simple fusion ways employed in CTRL including element-wise product, element-wise sum, and concatenation. MCF extends the visual/textual vector to the circulant matrix, which can fully exploit the interactions of the visual and textual representations. By plugging MCF into CTRL, the grounding accuracy is further improved.

Previous works like CTRL, ACRN and MCF directly calculate the visual-semantic correlation without explicitly modeling the activity information within two modalities, and the candidate segments fairly sampled by sliding window may contain various meaningless noisy contents that do not contain any activity. Hence, Ge et al. [38] explicitly mine activity concepts from both visual and textual parts as prior knowledge to provide an actionness score for each candidate segment, reflecting how confident it contains activities, which enhances the localization accuracy. MMRG [159] employs a multi-modal relational graph explicitly considering the interactions among visual and textual objects. It also designs customized pre-training tasks to enhance the visual representations.

Even though these sliding window-based techniques are straightforward and efficient, their computation is inefficient because the dense sampling process with predefined multi-scale sliding windows recalculates too many overlapped areas.

Proposal-Generated

Considering the inevitable drawbacks of sliding window-based methods, some approaches focus on reducing the number of proposal candidates, namely proposal-generated methods. These proposal-generated methods still adopt a two-stage scheme but avoid densely sampling through sliding windows, using different kinds of proposal networks.

QSPN [144] alleviates this computational burden by proposing temporal segments conditioned on the query, thereby reducing the number of candidate segments (c.f., Fig. 2.4). As shown in Fig. 2.4a, the query-guided SPN first incorporates the query embeddings into the video features to compute the attention weight for

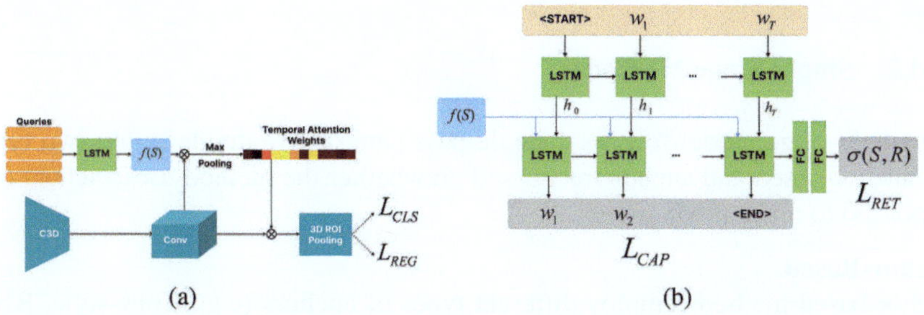

Fig. 2.4 The structure of QSPN, including the query-guided segment proposal network (c.f., **a**) and a fine-grained early-fused similarity model for retrieval (c.f., **b**)

each temporal location, and further integrates these temporal attention weights into the convolutional process for video encoding to propose query-aware representations of candidate segments. Afterwards, the generated proposal visual feature from Fig. 2.4a is incorporated into the sentence embedding process at each time step of the second layer of the two-layer LSTM in an early fusion way (c.f., Fig. 2.4b). QSPN devises a triplet-based retrieval loss which is similar to MCN:

$$\mathcal{L}_{RET} = \sum_{(S,R,R')} \max\{0,\ \eta + \sigma(S, R') - \sigma(S, R)\}, \tag{2.4}$$

where (S, R) is the positive (sentence, segment) pair while R' is the sampled negative segment. QSPN also devises an auxiliary captioning task that regenerates the query sentence from the retrieved video segment. The loss for captioning is as follows:

$$\mathcal{L}_{CAP} = -\frac{1}{KT}\sum_{k=1}^{K}\sum_{t=1}^{T_k} \log P(w_t^k | f(R), h_{t-1}^{(2)}, w_1^k, \ldots, w_{t-1}^k), \tag{2.5}$$

where a standard captioning loss is introduced to maximize the normalized log-likelihood of the words generated at all T unrolled time steps, over all K ground-truth matching query-moment pairs.

Similarly, SAP proposed by Chen and Jiang [13] integrates the semantic information of sentence queries into the generation process of activity proposals. Specifically, the visual concepts extracted from the query sentence and video frames are used to compute a visual-semantic correlation score for every frame. Activity proposals are generated by grouping frames with high visual-semantic correlation scores.

Summary Despite the intuitiveness and success of this two-stage matching-based paradigm, it also has some drawbacks. In order to achieve high localization accuracy (i.e., the candidate pool should have at least one proposal that is close to the ground-truth moment), the duration and location distributions of the candidate moments should be diverse, thus inevitably increasing the number of candidates, which leads to inefficient computation of the subsequent matching process.

2.1.1.2 Single-Stage Methods

The single-stage model follows a single-pass pattern. We divide it into two types, i.e., anchor-based and anchor-free, based on whether the method uses anchors (i.e., proposals) to make predictions.

Anchor-Based

Anchor-based methods employ different types of anchors (e.g., Conv-style, RNN-style) to generate candidate moments. TGN [17] adopts a typical single-stage deep

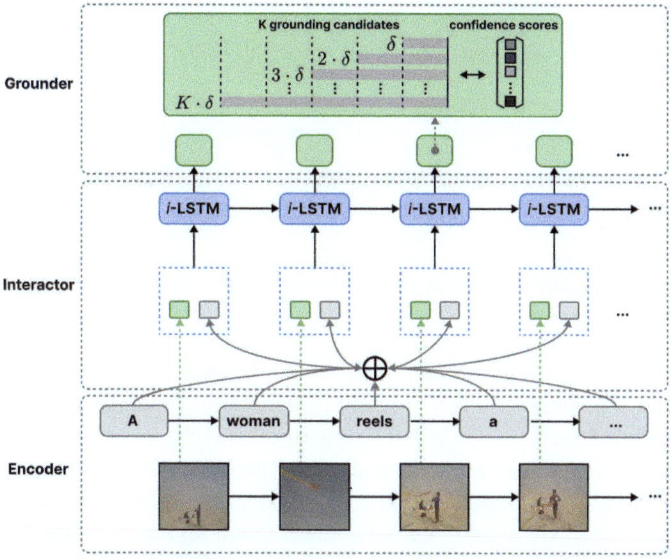

Fig. 2.5 The architecture of TGN, adopting a frame-by-word interaction single-stream framework. The grounder would generate multi-scale grounding candidates (anchors) that end at the same time step

architecture, which can localize the target moment in one single pass without handling heavily overlapped pre-segmented candidate moments. As shown in Fig. 2.5, TGN dynamically matches the sentence and video units via a sequential LSTM grounder with fine-grained frame-by-word interaction, and at each time step, the grounder would simultaneously score a group of candidate segments with different temporal scales ending at this time step.

CMIN [162] sequentially scores a set of candidate moments of multi-scale anchors like TGN but with a sequential BiGRU network, and refines the candidate moments with boundary regression. To further enhance the cross-modal matching, it devises a novel cross-modal interaction network, which first leverages a syntactic GCN to model the syntactic structure of queries, and captures long-range temporal dependencies of video context with a multi-head self-attention module.

Likewise, CBP [129] builds a single-stream model with sequential LSTM, which jointly predicts temporal anchors and boundaries at each time step to yield precise localization. To better detect semantic boundaries, CBP devises a self-attention-based module to collect contextual clues instead of simply concatenating the contextual features such as [36, 38, 45].

CSMGAN [79] also adopts a single-pass scheme. It builds a joint graph for modeling the cross-/self-modal relations via iterative message passing to capture the high-order interactions between two modalities effectively. Each node of the graph aggregates the messages from its neighbor nodes in an edge-weighted manner and updates its state with both the aggregated message and current state through

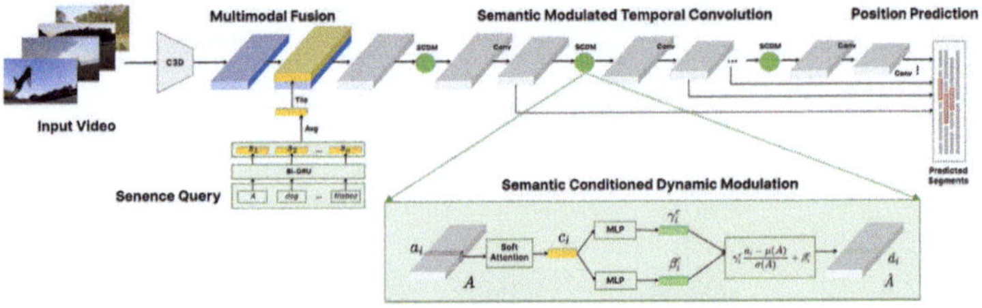

Fig. 2.6 The architecture of SCDM, which couples semantics-conditioned dynamic modulation into the temporal convolutional network to correlate the sentence-aware video contents over time

ConvGRU. Qu et al. [104] present a fine-grained iterative attention network (FIAN), which devises a content-oriented strategy to generate candidate moments different from the anchor-based methods with sequential RNNs mentioned above. FIAN employs a refined cross-modal guided attention block to capture the detailed cross-modal interactions, and further adopts a symmetrical iterative attention to generate both sentence-aware video and video-aware sentence representations.

While TGN establishes the temporal grounding architecture through a sequential LSTM network, Yuan et al. [155] propose SCDM, which exploits a hierarchical temporal convolutional network to conduct target segment localization, and couples it with a semantics-conditioned dynamic modulation to fully leverage sentence semantics to compose the sentence-related video contents over time. As shown in Fig. 2.6, the multi-modal fusion module fuses the entire sentence and each video clip in a fine-grained manner. The fused representation is formulated as:

$$\mathbf{f}_t = \mathrm{ReLU}(\mathbf{W}^f(\mathbf{v}_t||\bar{\mathbf{s}}) + \mathbf{b}^f)\,. \tag{2.6}$$

With such fused representations as inputs, the semantic modulated temporal convolution module further correlates sentence-related video contents in a temporal convolution procedure, dynamically modulating the temporal feature maps conditioned on the sentence. Specifically, for each temporal convolutional layer, the feature map is denoted as $\mathbf{A} = \{\mathbf{a}_i\}$. The feature unit $\mathbf{a}_i$ will be modulated based on the modulation vectors γ_i^c and β_i^c:

$$\hat{\mathbf{a}}_i = \gamma_i^c \cdot \frac{\mathbf{a}_i - \mu(\mathbf{A})}{\sigma(\mathbf{A})} + \beta_i^c\,, \tag{2.7}$$

where the modulation vectors are computed based on the sentence representation $\mathbf{S} = \{\mathbf{s}_n\}_{n=1}^N$:

$$\begin{aligned}\rho_i^n &= \mathrm{softmax}(\mathbf{w}^T\tanh(\mathbf{W}^s\mathbf{s}_n + \mathbf{W}^a\mathbf{a}_i + \mathbf{b}))\,,\ \mathbf{c}_i = \sum_{n=1}^{N}\rho_i^n\mathbf{s}_n\,,\\ \gamma_i^c &= \tanh(\mathbf{W}^\gamma\mathbf{c}_i + \mathbf{b}^\gamma)\,,\ \beta_i^c = \tanh(\mathbf{W}^\beta\mathbf{c}_i + \mathbf{b}^\beta)\,.\end{aligned} \tag{2.8}$$

Finally, the position prediction module outputs the location offsets and overlap scores of candidate video segments based on the modulated features. Similar to SCDM, RMN [78] also correlates video contents conditioned on the query semantics via a modulation module, and it further employs a cascade of several rectification-modulation layers for multi-step reasoning.

MAN [160] leverages a temporal convolutional neural network to address the *Video Grounding* task as well, where the sentence query is integrated as dynamic filters into the convolutional process. Specifically, MAN encodes the entire video stream using a hierarchical convolutional neural network to produce multi-scale candidate moment representations. The textual features are encoded as dynamic filters and convolved with such visual representations. Additionally, MAN exploits the graph-structured moment relation modeling adapted from Graph Convolution Network (GCN) [64] for temporal reasoning to further improve the moment representations. Similar to MAN, Soldan et al. [113] also adopt GCN and present a video-language graph matching network (VLG-Net) for modeling the fine-grained inter-modal interaction.

Both SCDM and MAN only consider 1D temporal feature maps, while the 2D-TAN [165] network models the temporal relations of video segments via a two-dimensional map. As shown in Fig. 2.7, it first divides the video into evenly spaced video clips with duration τ. The (i, j)-th location on the 2D temporal map represents a candidate moment (or anchor) from the time $i\tau$ to $(j+1)\tau$. This kind of 2D temporal map covers diverse video moments with different lengths while representing their adjacent relations. The proposed temporal adjacent network fuses the sentence representation with each of the candidate moment features and then leverages a convolutional neural network to embed the video context information, and finally predicts the confidence score of each candidate to be the final target segment.

Some works [12, 34, 130, 171] also adopt the same proposal generation approach as that of 2D-TAN. A structured multi-level interaction network (SMIN) is proposed by Wang et al. [130], which further modifies the 2D temporal feature map as its proposal generation module. SMRN [12] adds a residual connection within the hier-

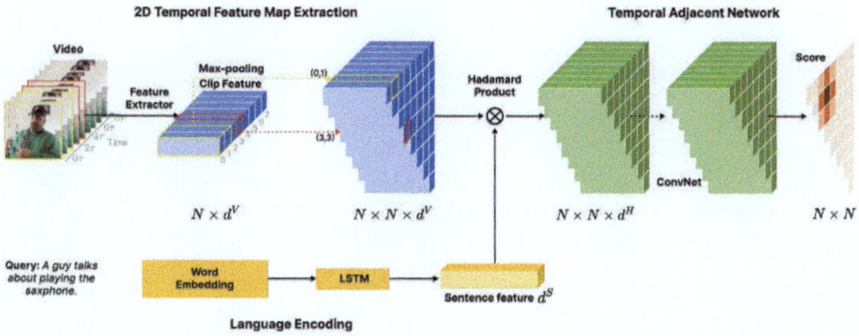

Fig. 2.7 The architecture of 2D-TAN, which consists of a text encoder for language representation, a 2D temporal feature map extractor for video representation and a temporal adjacent network for moment localization

archical convolution network of 2D-TAN and further utilizes the query semantics to modulate short connections within residual blocks. Zhang et al. [171] present a visual-language transformer backbone followed by a multi-stage aggregation module to get discriminative moment representations for more accurate localization. Gao et al. [34] design a fine-grained semantic distillation framework for retrieving desired moments with superiority in both accuracy and efficiency.

It is important to highlight that Bao et al. [4] introduce an anchor-based dense event propagation network (DepNet) designed for the more challenging task of dense events grounding. This task aims to localize multiple temporal moments based on a given paragraph. DepNet effectively aggregates the visual-semantic information from dense events into a compact representation and subsequently propagates this information to accurately localize individual events, thereby fully leveraging the temporal relationships between these events.

Despite the superior performance anchor-based methods have achieved, the performance is sensitive to the heuristic rules manually designed (i.e., the number and scales of anchors). As a result, such anchor-based techniques are unable to generalize to situations where the length of the videos varies. Furthermore, while pre-segmentation, such as that employed in two-stage methods, is not necessary, the approach still fundamentally relies on the ranking of proposal candidates, which can impact its efficiency.

Anchor-Free

Instead of ranking a vast number of proposal candidates, the anchor-free methods start from more fine-grained video units such as frames or clips, and aim to predict the probability for each frame/clip being the start and end point of the target segment, or directly regress the start and end points from the global view.

Yuan et al. [156] propose ABLR, which solves *Video Grounding* from a global perspective without generating anchors. Specifically, as shown in Fig. 2.8, to preserve the context information, ABLR first encodes both video and sentence via bidirectional LSTM networks. Then, a multi-modal co-attention mechanism is introduced to generate both video attention which reflects the global video structure and sentence attention, highlighting the crucial details for temporal localization.

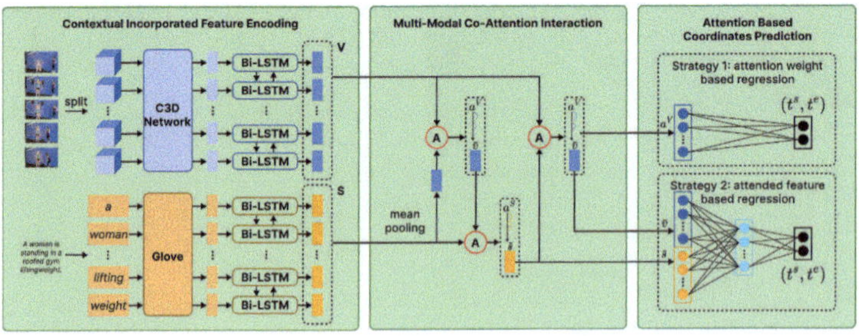

Fig. 2.8 The architecture of ABLR, which regresses the target coordinates with a multi-modal co-attention mechanism

Finally, an attention-based coordinate prediction module is designed to regress the temporal coordinates (i.e. the starting timestamp t^s and the ending timestamp t^e) of sentence query from the former output attentions. Meanwhile, there are two different regression strategies (i.e., attention weight-based regression and attended feature-based regression) with the location regression loss $L_{reg}^{ablr} = \sum_{i=1}^{K}[R(\tilde{t}_i^s - t_i^s) + R(\tilde{t}_i^e - t_i^e)]$, where R is a smooth L1 function. In addition to the location regression loss, which aims to minimize the distance between the temporal coordinates of the predicted segments and the ground truth segments, ABLR also introduces an attention calibration loss $\mathcal{L}_{cal}$ to enhance the accuracy of video attention mechanisms.

$$\mathcal{L}_{cal} = -\sum_{i=1}^{K} \frac{\sum_{j=1}^{M} m_{i,j} \log(a_j^{V_i})}{\sum_{j=1}^{M} m_{i,j}} . \tag{2.9}$$

Here, $\mathcal{L}_{cal}$ encourages the attention weights of the video clips within the ground truth segment to be higher.

LGI [97] formulates the *Video Grounding* task as an attention-based location regression like ABLR. It further presents a more effective local-global video-text interaction module, which models the multi-level interactions between semantic phrases and video segments. Chen et al. [21] propose pairwise modality interaction (PMI) via a channel-gated modality interaction model to explicitly model the channel-level and sequence-level interactions in a pairwise fashion. Specifically, a light-weight convolutional neural network is applied as the localization head to process the feature sequence and output the video-text relevance score and boundary prediction. HVTG [14] also computes the frame-level relevance scores and makes boundary prediction based on these scores. To perform the fine-grained interaction among the visual objects and between the visual object and the language query, HVTG devises a hierarchical visual-textual graph to encode the features.

Unlike ABLR which regresses the coordinates of the target moment directly, ExCL [39] borrows the idea from the Reading Comprehension task [16] in the natural language processing area. The process of retrieving a video segment from the video is analogous to extracting a text span from the passage. Specifically, as shown in Fig. 2.9, ExCL employs three different variants of start-end frame predictor networks (i.e., MLP, Tied-LSTM and Conditioned-LSTM) to predict start and end probabilities for each frame. It has two training modes that depend on the training objective. ExCL-clf uses a classification loss, which is trained with negative log-likelihood loss:

$$L(\theta) = -\frac{1}{N}\sum_{i}^{N} \log(P_{\text{start}}(t_s^i)) + \log(P_{\text{end}}(t_e^i)) , \tag{2.10}$$

While ExCL-reg trains using a regression loss, it computes an expectation over the probability distribution provided by SoftMax outputs to predict start and end times:

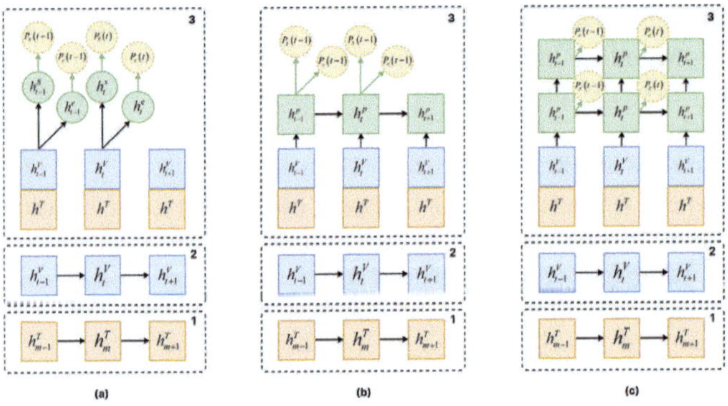

Fig. 2.9 The architecture of ExCL. It consists of three modules: a sentence encoder (shown in orange squares), a video encoder (shown in blue squares) and three variants of frame predictor (i.e., (**a**) MLP, (**b**) Tied-LSTM and (**c**) conditioned-LSTM). The frame predictor outputs the start and end probabilities for each frame

$$
\begin{aligned}
t_s &= \mathbb{E}_{P_{\text{start}}}[t] = \sum_{t_s=1}^{T} t_s P_{\text{start}}(t_s)\,, \\
t_e &= \mathbb{E}_{P_{\text{start}}}[\mathbb{E}_{P_{\text{end|start}}}[t]] = \sum_{t_s=1}^{T} P_{\text{start}}(t_s) \sum_{t_e=1}^{T} P_{\text{end|start}}(t_e)\,, \\
P_{\text{end|start}} &= \text{SoftMax}(\mathbb{1}[t_e \geq t_s] S_{\text{end}}(t))\,.
\end{aligned}
\tag{2.11}
$$

Likewise, VSLNet [164] employs a standard span-based Question Answering framework. VSLNet further distinguishes the differences between video sequence and text passage for better usability for *Video Grounding* task. To address the differences, it designs a query-guided highlighting strategy to narrow down the search space to a smaller coarse highlight region. L-Net [18] presents a boundary model that uses the language query to semantically localize the video segment by predicting the start and end boundaries. It devises a cross-gated attended recurrent network to emphasize the relevant video parts while the irrelevant ones are gated out, and a cross-modal interactor for fine-grained interactions between two modalities.

TMLGA [107] also predicts start and end probabilities for each video unit. It further models the uncertainty of boundary labels, using two Gaussian distributions as ground truth probability distributions. CPN [174] devises a cascaded prediction network based on the segment-tree data structure. It performs two sub-tasks (i.e., decision navigation and signal decomposition) on each level from top to bottom for final boundary prediction. PEARL [172] integrates the subtitles of videos and convolves the query filters into the visual and subtitle branches to locate the boundaries.

Lu et al. [84] propose a dense bottom-up grounding framework (DEBUG), which localizes the target segment by predicting the distances to bidirectional temporal boundaries for all frames inside the ground truth segment. By adopting this approach, the significant imbalance issue caused by considering only the boundaries of the ground truth segment as positive samples is effectively addressed. Consequently, all frames within the ground truth segment can be treated as positive samples. A typical dense anchor-free model generally comprises a backbone framework for multi-modal feature encoding and a head network designed for frame-level predictions. Specifically, DEBUG adopts QANet as its backbone network which models the interaction between videos and queries, and designs three branches as head networks which aim to separately predict the classification score, boundary distances, and confidence score for each frame.

Similarly, DRN [158] and GDP [20] also adopt such a dense anchor-free framework. For the backbone, DRN uses a video-query interaction module to obtain fused hierarchical feature maps. For the head network, DRN densely predicts the distances to boundaries, matching score and estimated IoU for each frame within the ground truth segment. Meanwhile, for the backbone, GDP leverages a Graph-FPN layer which conducts graph convolution over all nodes in the scene space to enhance the integrated frame features. For the head network, GDP predicts the distances from its location to the boundaries of the target moment and a confidence score to rank its boundary prediction for each frame.

Another graph-based method DORi [108] utilizes a spatio-temporal graph to model the temporally-changing inter-object interactive relationships based on the language query, which can further improve the activity representations. Instead of adopting well-designed graph-based structures, Yu et al. [154] propose a simple yet effective approach that only conducts bi-directional cross-modal interaction via multi-head attention with multiple training objectives.

Different from the aforementioned single-stage methods which either sample from multi-scale anchors or directly regress the final coordinates, some methods out of these patterns have emerged. The boundary proposal network (BPNet) [142] keeps the advantages of both anchor-based and anchor-free methods and avoids the defects, which generates proposals by anchor-free methods and then matches them with the sentence query in an anchor-based manner. Wang et al. [128] propose a dual path interaction network (DPIN) containing two branches (i.e., a boundary prediction pathway for frame-level features and an alignment pathway for segment-level features) to localize the target moment in a complementary manner. Inspired by the dependency tree parsing task in the natural language processing area, a biaffine-based architecture named context-aware biaffine localizing network (CBLN) [80] is proposed which can simultaneously score all possible pairs of start and end indices. Ding et al. [28] introduce a support-set cross-supervision (Sscs) module. The Sscs module can be a plug-in branch to enhance multi-modal relation modeling for both anchor-based and anchor-free methods.

We then present the DETR series of models, which have been thoroughly studied and have demonstrated excellent efficacy. DETR (Detection Transformer) [11] is a pioneering Transformer-based object detection model. Unlike traditional object detection methods, it eliminates the need for non-maximum suppression (NMS)

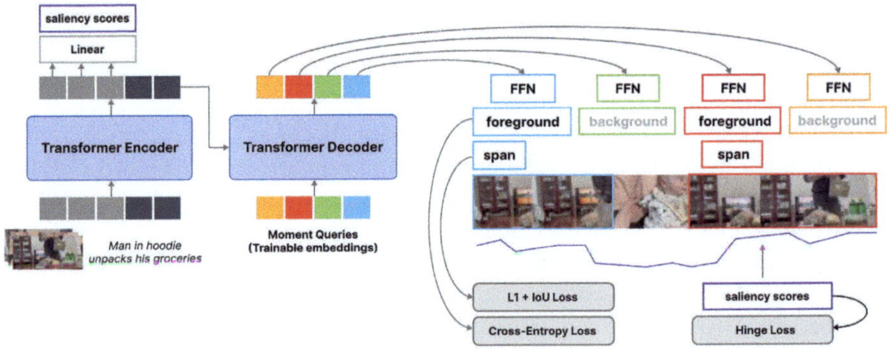

Fig. 2.10 The architecture of Moment-DETR. It consists of a transformer encoder-decoder and three prediction heads for predicting saliency scores, fore-/back-ground scores and moment coordinates

post-processing steps and anchor-based priors, allowing for end-to-end training. This simplification significantly streamlines the object detection pipeline.

Building upon DETR, Lei et al. [69] developed Moment-DETR, designed for joint video grounding and highlight detection tasks. Moment-DETR is inspired by DETR's architecture, framing video grounding as a set prediction problem. It views video grounding as an end-to-end task and simplifies the overall workflow. Specifically, Moment-DETR employs a two-layer perceptron to project both video and query features into a shared embedding space. As shown in Fig. 2.10, the concatenated features are fed into a Transformer encoder, which processes the input sequence using multiple layers of self-attention and feed-forward networks. The decoder part, similar to DETR, uses moment queries to interact with encoder outputs, predicting both the temporal location (center and width) of moments in the video and their relevance. Lei et al. argue that a single query may correspond to multiple segments in a video. As a result, both the predicted moments and the ground-truth moments are treated as sets. The Hungarian algorithm is used to match these two sets optimally, allowing for a more accurate computation of the loss. The final loss includes four components: saliency loss, cross-entropy loss (to classify segments as foreground or background), and two losses measuring the difference between predicted and ground-truth moments (L1 loss and generalized IoU loss [106]). To enhance performance, Lei et al. also experimented with weakly supervised pre-training using Automatic Speech Recognition (ASR) captions. These noisy captions were used to predict timestamps, further improving the model's ability to handle video grounding and highlight detection tasks.

Moment-DETR has notable limitations, such as assuming that a text query always exists and being limited to only the visual modality of a video. While serving as a strong baseline, it still offers basic functionality. UMT [82] addresses these issues by exploring multi-modal learning and flexibility. It introduces a unified framework that incorporates not only video and text but also audio as important inputs. UMT is designed to handle varying modality reliability, adapting to situations where one modality, such as text or audio, may be unreliable or absent. For instance, if text input is unavailable, UMT focuses on highlight detection. Through its flexible architecture, UMT is able to accommodate various combinations of input modalities without requiring specifically designed models for each variation.

While Moment-DETR employs an efficient transformer encoder-decoder architecture, it struggles with properly utilizing global context and effective cross-modal interactions. The model naively concatenates visual and textual features, weakening their correlation. MH-DETR [147] builds upon UMT by reintroducing key components like the moment decoder and bipartite matching, which UMT had removed. It enhances intra-modal context through a simple pooling operator within the encoder and integrates cross-modal interaction to better align visual and textual features. This allows MH-DETR to provide more precise predictions for video grounding by focusing on the visual elements most relevant to the text query.

The intricacy of videos is not well captured by moment queries in earlier models, such as Moment-DETR, which frequently treat them as fixed input-agnostic regions. EaTR [54] addresses this by introducing dynamic moment queries that correspond to event units within the video. This model performs two levels of reasoning: event reasoning, which identifies distinct events in the video, and moment reasoning, which fuses the moment queries with the given sentence to predict the final timestamps. By adopting a slot attention mechanism, EaTR dynamically identifies the most relevant event units, minimizing irrelevant information and improving convergence speed. A gated fusion transformer layer further enhances accuracy by ensuring that sentence information is properly aggregated into the moment queries.

Previous models have not effectively tackled the problem of weak alignment in video moment detection. BM-DETR [61] focuses on this issue by utilizing both positive and negative queries within its transformer-based encoder-decoder architecture. By calculating joint probabilities for frames based on positive and negative queries, BM-DETR improves the true positive rate of moment detection. In order to improve semantic alignment between the text and video queries, the model also makes use of cross-modal discrimination. Additionally, a temporal shifting mechanism improves the robustness of the model, ensuring better performance in capturing contextual meaning throughout the video.

While previous models advanced moment retrieval (MR) and highlight detection (HD), they overlook the critical role of the text query in aligning video clips with their corresponding moments. QD-DETR [95] directly addresses this by ensuring that the prediction for each clip is highly dependent on the query. It introduces cross-attention layers early in the transformer encoder to tightly integrate textual and visual features. Furthermore, it generates negative video-query pairs to train the model to suppress irrelevant pairs, increasing the contribution of the text query in the final prediction. QD-DETR produces query-dependent video representations, ensuring that the model's performance is consistently tied to the relevance of the text query in both MR and HD tasks.

Previous methods often treat Moment Retrieval (MR) and Highlight Detection (HD) as a simple multi-task problem, optimizing both tasks together using generic strategies like transformer-based models. However, this approach overlooks the intrinsic differences between these tasks, leading to inconsistent performance in moment localization and highlight estimation. MR requires understanding local relationships within a video, while HD focuses on global saliency across the entire video. To address this, UVCOM [143] introduces a unified framework that balances both local and global requirements. It leverages a Comprehensive Integration Module (CIM) to progressively fuse intra- and inter-modal information, allowing for more accurate local moment detection and global highlight estimation. Additionally, UVCOM employs multi-aspect contrastive learning to strengthen the relationship between clip-text alignment and video-linguistic discrimination, thereby enhancing performance in both tasks.

Conventional joint models typically use a shared backbone to learn multi-modal features for both MR and HD tasks, assuming a high correlation between them. However, this bottom-up, data-driven approach tends to oversimplify the relationship between MR and HD by relying on common features and neglecting their task-specific characteristics. TaskWeave [151] proposes a task-driven paradigm to overcome these limitations by considering the distinct objectives of MR and HD. It introduces a task-decoupled unit with a shared expert for common feature extraction and task-specific experts for unique characteristics. Furthermore, an inter-task feedback mechanism allows MR and HD predictions to influence each other, improving overall task performance. The model also employs a dynamic task-dependent loss, automatically adjusting weights to optimize both tasks jointly.

While previous models like QD-DETR improve cross-attention for text reflection, they fail to differentiate the degree of text engagement across video segments. CG-DETR [94] addresses this limitation by introducing Adaptive Cross-Attention (ACA), which adjusts text-to-video attention based on the relevance between the video and text query. This model focuses on the clip-wise correlation between textual queries and video segments, refining the attention map to highlight relevant moments. Additionally, a moment-adaptive saliency detector leverages instance-specific context to predict the saliency of each video clip, enhancing the model's capacity to more accurately identify significant video segments.

Most query-based models assume that a moment's center is the best representation for the associated sentence, which can lead to misaligned predictions. BAM-DETR [67] challenges this assumption by proposing a boundary-oriented approach that uses an anchor point and the distances to the boundaries, rather than relying on a moment's center. This allows for more flexible and accurate boundary detection. The model adopts a dual-pathway decoding process: one for identifying anchor points and another for refining boundaries. This approach, combined with a localization-oriented ranking system, enhances the model's ability to predict more precise temporal boundaries for grounded moments.

Traditional joint models for MR and HD tend to ignore the reciprocal relationship between the two tasks, missing opportunities to boost performance. TR-DETR [120] takes advantage of this interdependence by introducing a task cooperation module. In this model, highlight scores from HD help filter irrelevant clips in MR, while MR outcomes inform HD's understanding of user queries. In order to improve the model's joint task performance, the framework employs a local-global multi-modal alignment module to fine-tune feature interactions and guarantee more precise moment retrieval and highlight detection.

Previous approaches often suffer from generating overlapping and redundant proposals due to a lack of task-specific guidance. RGTR [121] addresses this by introducing regional priors to guide the query process for moment retrieval. The model utilizes anchor pairs as moment queries, which are initialized with regional priors to enhance query diversity. This allows RGTR to better differentiate between moments and reduce redundant predictions. Additionally, the model incorporates an IoU-aware scoring head that takes both classification confidence and boundary quality into account, leading to more accurate localization.

In summary, the DETR-based models have significantly advanced the tasks of moment retrieval and highlight detection by leveraging Transformer architectures for end-to-end learning. Moment-DETR is the foundational model, simplifying video grounding through set prediction and cross-modal interactions, but faces limitations like fixed modality reliance. Models like UMT and MH-DETR expand this by incorporating multi-modal inputs and improving cross-modal attention, enhancing prediction accuracy and flexibility. Further innovations include EaTR's dynamic moment queries for better event reasoning, BM-DETR's positive-negative query system for more precise moment alignment, and QD-DETR's query-dependent design, which strengthens text-to-video alignment. Advanced models like UVCOM and TaskWeave introduced task-specific mechanisms to balance the distinct requirements of MR and HD, optimizing both tasks simultaneously. Lastly, models such as CG-DETR, BAM-DETR, and TR-DETR refine attention mechanisms and boundary detection, addressing challenges like misalignment and overlapping proposals, while RGTR introduced regional priors for more accurate moment localization. Overall, the evolution of DETR-based models has significantly improved the precision, flexibility, and effectiveness of both MR and HD tasks by addressing critical challenges in cross-modal interaction and task differentiation.

2.1.1.3 Reinforcement Learning-Based Methods

As another kind of anchor-free approach, RL-based frameworks view such a task as a sequential decision process. The action space for each step is a set of handcrafted-designed temporal transformations (e.g., shifting, scaling).

He et al. [44] first introduce deep reinforcement learning techniques to address the task of *Video Grounding*, which formulates *Video Grounding* as a sequential decision-making problem. As depicted in Fig. 2.11, at each time step, the observation network outputs the current state of the environment for the actor-critic module to generate an action policy (i.e., the probabilistic distribution of all the actions predefined in the action space), based on which the agent will perform an action to adjust the temporal boundaries. This iterative process will be ended when encountering the STOP action or reaching the maximum number of steps (i.e., T_{max}). Specifically, at each step, the current state vector is computed as

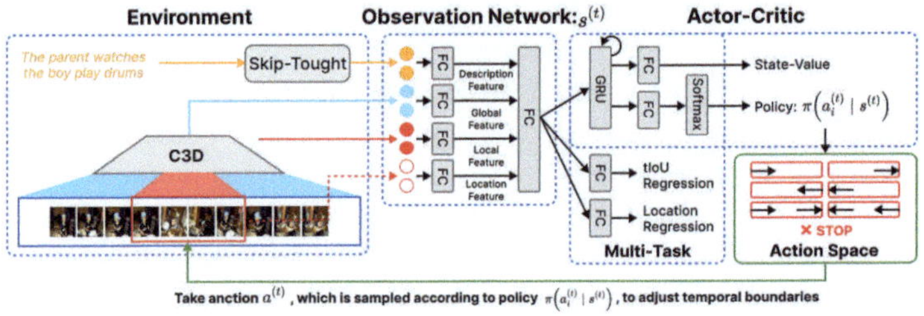

Fig. 2.11 The architecture of R-W-M framework. The observation network takes the environment (i.e., description feature, global video feature, local feature and location feature) as input to compute the current state. Then one of those seven operators in the action space is determined to adjust the temporal boundaries of the current segment. Two auxiliary supervised tasks including tIoU regression and location regression are also leveraged

$s^{(t)} = \Phi(E, V_G, V_L^{(t-1)}, L^{(t-1)})$, where $s^{(t)}$ is generated by one FC layer whose inputs are the concatenated features including the segment-specific features (i.e., the normalized boundary pair $L^{(t-1)} = [l_s^{(t-1)}, l_e^{(t-1)}]$ and local segment C3D feature $V_L^{(t-1)}$) and global features (i.e., the sentence embedding E and entire video C3D feature V_G). Then the actor-critic module employs GRU to model the sequential decision-making process. At each time step, GRU takes $s^{(t)}$ as input and the hidden state is used for policy (denoted as $\pi(a_i^{(t)}|s^{(t)}, \theta_\pi)$) generation and state-value (denoted as $v(s^{(t)}|\theta_v)$) estimation. The reward for each step r_t is designed to encourage a higher tIoU compared to that of the last step. The accumulated reward function is then defined as (γ is a constant discount factor):

$$R_t = \begin{cases} r_t + \gamma * v(s^{(t)}|\theta_v), & t = T_{max} \\ r_t + \gamma * R_{t+1}, & t = 1, 2, \ldots, T_{max} - 1 \end{cases} . \tag{2.12}$$

Then they introduce the advantage function as objective which is approximated by the Mente Carlo sampling to get the policy gradient:

$$\mathcal{L}'_A(\theta_\pi) = -\sum_t (\log \pi(a_i^{(t)}|s^{(t)}, \theta_\pi))(R_t - v(s^{(t)}|\theta_v)) . \tag{2.13}$$

They further leverage two supervised tasks (i.e., tIoU regression and location regression) so the parameters can be updated from both policy gradient and supervised gradient to help the agent obtain more accurate information about the environment.

Wang et al. [127] propose an RNN-based RL model which sequentially observes a selective set of video frames and finally obtains the temporal boundaries given the query. Cao et al. [7] firstly leverage the spatial scene tracking task, which utilizes a spatial-level RL for filtering out the information that is not relevant to the text query. The spatial-level RL can enhance the temporal-level RL for adjusting the temporal boundaries of the video. TripNet [42] uses gated attention to align textual and visual features, leading to improved accuracy. It incorporates a policy network for efficient search, which selects a fixed temporal bounding box moving around without watching the entire video.

In contrast to traditional RL-based techniques, TSP-PRL [141] uses a tree-structured policy that draws inspiration from the coarse-to-fine decision-making paradigm of humans. The agent receives the state from the environment (video clips) and estimates a primitive action via tree-structured policy, including root policy and leaf policy. The action selection is depicted by a switch over the interface in the tree-structured policy. The alignment network will predict a confidence score to determine when to stop. Meanwhile, AVMR [8] addresses *Video Grounding* under the adversarial learning paradigm, which designs an RL-based proposal generator to generate proposal candidates and employs Bayesian Personalized Ranking as a discriminator to rank these generated moment proposals in a pairwise manner.

2.1.2 Weakly Supervised Methods

When annotating the ground truth data for *Video Grounding*, the annotators should read the query and watch the video first, and then determine the start and end points of the query-indicated segment in the video. Such a human-labored process is very time-consuming. Therefore, due to the labor-intensive ground truth annotation procedure, some works start to extend *Video Grounding* to a weakly supervised scenario where the locations of ground truth segments (i.e., the start and end timestamps) are unavailable in the training stage. This is formally named as weakly supervised *Video Grounding*. In general, weakly supervised methods for *Video Grounding* can be grouped into two categories (i.e., MIL-based and reconstruction-based). One representative work will be illustrated in detail for each category, after which we will introduce the remaining.

2.1.2.1 MIL-based Methods

Some works [23, 37, 93, 122] adopt multi-instance learning (MIL) to address the weakly *Video Grounding* task. When temporal annotations are not available, the whole video is treated as a bag of instances with bag-level annotations, and the predictions for instances (video segment proposals) are aggregated as the bag-level prediction.

TGA [93] is a typical MIL-based method which learns the visual-text alignment in the video level by maximizing the matching scores of the videos and their corresponding descriptions while minimizing the matching scores of the videos and the descriptions of others. It presents text-guided attention (TGA) to get text-specific global video representations, learning the joint representation of both the video and the video-level description. As illustrated in Fig. 2.12, TGA first employs a GRU

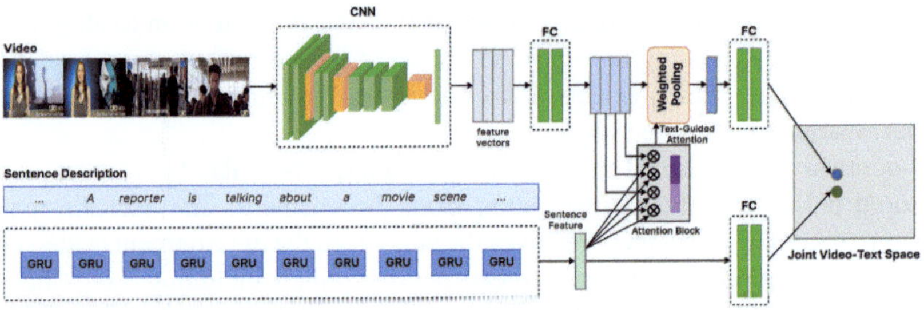

Fig. 2.12 The overall framework of TGA. It learns a joint embedding network to align the text and video features. The global video representation is generated by weighted pooling based on multiple text-guided attention layers

for sentence embedding and a pre-trained image encoder for extracting frame-level features. The similarity between jth sentence and the kth temporal feature within the ith video denoted as s^i_{kj} is computed and a softmax operation is applied to get the text-guided attention weights for each temporal unit denoted as a^i_{kj}:

$$s^i_{kj} = \frac{\mathbf{w}^{i\,T}_j \mathbf{v}^i_k}{\left\|\mathbf{w}^i_j\right\|_2 \left\|\mathbf{v}^i_k\right\|_2}, \quad a^i_{kj} = \frac{\exp(s^i_{kj})}{\sum_{m=1}^{nv_i} \exp(s^i_{mj})}. \tag{2.14}$$

Thus we could get the sentence-wise global video feature $\mathbf{f}^i_j = \sum_{k=1}^{nv_i} a^i_{kj} \mathbf{v}^i_k$. Then, the video feature vector and paired text feature vector are projected into the joint embedding space. A hinge loss is used to learn a joint embedding such that the positive pairs are closer than the negative pairs in the feature space.

WSLLN [37] is another MIL-based end-to-end weakly supervised language localization network conducting clip-sentence alignment and segment selection simultaneously. Huang et al. [51] present a cross-sentence relations mining (CRM) method exploring the cross-sentence relations within paragraph-level scope to improve the per-sentence localization accuracy.

A video-language alignment network (VLANet) proposed by Ma et al. [90] prunes the irrelevant moment candidates with the Surrogate Proposal Module and utilizes multi-directional attention to get a sharper attention map for better multi-modal alignment. In an attempt to apply an RL-based model for weakly *Video Grounding*, Wu et al. [140] suggest a boundary adaptive refinement framework (BAR) achieving grounding results that are both content-aware and boundary-flexible. Chen et al. [23] propose a novel coarse-to-fine model (WSTG) based on MIL. The coarse stage first chooses a rough passage that semantically matches the provided sentence from a collection of predefined sliding windows. Afterwards, the fine stage mines the fine-grained matching relationship between each frame in the coarse segment and the sentence. It thereby refines the boundary of the coarse segment by grouping the frames and gets more precise grounding results.

Tan et al. [122] propose a Latent Graph Co-Attention Network (LoGAN), a novel co-attention model that performs fine-grained semantic reasoning over an entire video. LoGAN is also a MIL-based method, which performs a similar frame-by-word interaction with the supervised method TGN [17] and adapts the graph module from another supervised method MAN [160] for iterative frame representation update. Wang et al. [131] present a fine-grained semantic alignment network (FSAN), which enables iterative multi-head attention-based cross-modal interaction to capture fine-grained video-language alignment. In order to learn more robust and discriminative moment features, VCA [132] devises a visual co-occurrence alignment NCE loss that maximizes the similarity between video moments from various videos with similar descriptions.

Since MIL-based methods typically learn the visual-text alignment with a triplet loss, these methods heavily depend on the quality of randomly-selected negative samples, which are often easy to distinguish from the positive ones and cannot provide strong supervision signals.

2.1.2.2 Reconstruction-Based Methods

The reconstruction-based methods [15, 31, 72, 116] attempt to reconstruct the given sentence query based on the selected video segments and use the intermediate results for sentence localization. Unlike MIL-based methods, the reconstruction-based methods learn the visual-textual alignment in an indirect way. Lin et al. [72] propose a semantic completion network (SCN) to predict masked key terms within queries based on the visual context provided by generated and selected video proposals. Specifically, for each proposal G^k, denoted by $\hat{\mathbf{v}}^k = \{\mathbf{v}_i\}_{i=s_k}^{e_k}$, with the masked query representation $\hat{q}$, the energy word distribution $\mathbf{e}_i^k$ at i^{th} time step can be computed as $\mathbf{e}_i^k = \mathbf{W}_v\mathbf{f}_i^k + \mathbf{b}_v$, where $\mathbf{f}^k = \{\mathbf{f}_i^k\}_{i=1}^{n_q}$ are the cross-modal semantic representations, computed by $\mathbf{f}^k = \mathbf{Dec}_q(\hat{\mathbf{q}},\ \mathbf{Enc}_v(\hat{\mathbf{v}}^k))$, $\mathbf{Dec}_q$ and $\mathbf{Enc}_v$ are respectively the textual decoder and visual encoder based on bi-directional Transformer [126]. Afterwards, the reconstruction loss can be computed by adding up all negative log-likelihood of masked words:

$$\mathcal{L}_{rec}^k = -\sum_{i=1}^{n_q-1} \log p(\mathbf{w}_{i+1}|\hat{\mathbf{w}}_{1:i}, \hat{\mathbf{v}}^k) = -\sum_{i=1}^{n_q-1} \log p(\mathbf{w}_{i+1}|\mathbf{e}_i^k)\,. \tag{2.15}$$

Song et al. [116] present a Multi-Level Attentional Reconstruction Network (MARN), which leverages the idea of attentional reconstruction. MARN uses proposal-level attention layers to rank the segment candidates and refine them with clip-level attention layers.

Duan et al. [31] formulate and address the problem of weakly supervised dense event captioning in videos (i.e., to detect and describe all events of interest in a video), which is a dual problem of weakly supervised *Video Grounding*. It presents a cycle system to train the model which can solve such a pair of dual problems at the same time. In other words, weakly supervised *Video Grounding* can be regarded as an intermediate task in such a cycle system. Similar to Duan et al.'s work [31], Chen and Jiang [15] also employ a loop system for dense event captioning. They adopt a concept learner to construct an induced set of concept features to enhance the information passing between the sentence localizer and event captioner.

Chen et al. [24] tackle the limitations of previous weakly supervised video grounding methods by introducing a video composition strategy. Their approach models both inter-contrast and intra-contrast relationships among clips derived from diverse videos, thereby generating pseudo-labels for temporal boundaries. Through the composition of video pairs and the segregation of text-related segments from unrelated ones, the model effectively learns to ground sentences under weak supervision. Additionally, they augment performance by incorporating a transformer-based architecture and negative query samples, which ensures more accurate grounding without dependence on simplistic scene transitions.

CNM [175] addresses two key limitations in weakly supervised video grounding: the inadequate handling of negative samples from within the same video and the inefficiency of existing proposal generation methods. The majority of earlier methods concentrated on differentiating videos based on the semantic alignment between query sentences and video-level segments, but they failed to recognize the more difficult-to-identify mismatched segments within a single video. Additionally, proposal generation was often independent of the specific video and query, leading to a lack of flexibility and precision. As shown in Fig. 2.13, CNM solves these issues by dynamically generating informative proposals through a learnable Gaussian mask, which highlights relevant video segments. It also introduces an Intra-Video Contrastive (IVC) Loss to better handle negative samples within the same video, improving temporal localization performance.

Building on the limitations of CNM, CPL [176] aims to better handle long videos with multiple events, where the simple Gaussian mask approach of CNM may fail. While CNM uses a single mask for positive and negative samples, CPL introduces multiple learnable Gaussian functions to generate content-dependent proposals,

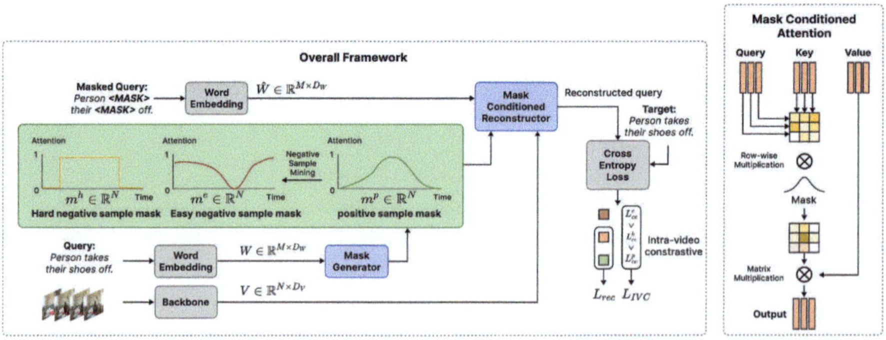

Fig. 2.13 The architecture of CNM

both positive and negative, capturing the complexity of the video. Moreover, it refines the negative proposal mining strategy by dynamically adjusting the difficulty of negative samples, starting with easier-to-distinguish segments and progressing to more confusing ones. This dynamic curriculum reduces ambiguity and optimizes the model for more reliable intra-video comparisons.

CPI [65] addresses the limitations of CNM and CPL, particularly their dependence on word-level masks, which can introduce unnecessary noise and diminish semantic understanding. CPI introduces a novel contrastive learning framework that utilizes pseudo-samples to dynamically intervene in the training process. This approach facilitates a more comprehensive understanding of video-text semantics by incorporating reverse inference, where the model predicts masked words based on video content. By balancing the use of masks with pseudo-sample-driven interventions, CPI enhances contrastive learning, especially in handling ambiguous visual scenes, thereby improving the alignment between multi-modal features.

While CNM and CPL improve proposal generation, they impose constraints such as the reliance on Gaussian masks, which can be too rigid. IRON [10] eliminates these constraints through iterative refinement of proposals without relying on predefined assumptions. It tackles two primary challenges: the absence of explicit correspondence modeling and the incomplete coverage of complex events in videos. IRON incorporates semantic and conceptual distillation to ensure that proposals are closely aligned with the query, while also enhancing proposal confidence scores via a label propagation technique. This iterative process generates more comprehensive and precise temporal intervals, capturing the entire event rather than merely its most prominent aspects.

SCANet [153] addresses the issue of "scene-proposal mismatch," where current methods inaccurately generate a fixed number of proposals regardless of the video's scene complexity. SCANet introduces a Scene Complexity-Aware Network that adapts the number of proposals based on the difficulty of detecting scenes within the video. It incorporates Complexity-Adaptive Proposal Generation (CPG) and Complexity-Adaptive Proposal Enhancement (CPE) to generate and improve proposals according to the scene complexity. Additionally, it mines hard negatives from the video and the larger video corpus by utilizing contrastive learning.

CCR [89] focuses on countering spurious correlations in cross-modality interaction when reconstructing masked queries. Traditional methods rely on similarities between video proposals and queries, but CCR disentangles the causal effects of language knowledge and cross-modality interactions. By introducing counterfactual reasoning, CCR minimizes the unimodal language bias during reconstruction, ensuring that predictions rely primarily on the interaction between video and masked queries. This approach enhances the quality of cross-modal alignment without being distorted by dataset biases.

UGS [52] leverages self-training to improve weakly supervised video grounding by introducing extra supervision signals. Unlike traditional methods that depend heavily on negative samples, UGS applies a Bayesian teacher network to estimate uncertainty, reducing error accumulation. Through cyclic mutual learning between student and teacher networks, UGS enhances temporal consistency and prevents overfitting. This method provides better quality control for supervision and improves performance across various datasets.

Previous methods often rely on single Gaussian proposals, which lack flexibility in representing complex and diverse events in a video. PPS [63] addresses this limitation by introducing a Gaussian Mixture Proposal (GMP) that learns the importance, centroid, and range of each Gaussian in the mixture to represent arbitrary shapes. A Pull-Push Scheme (PPS) is used to adjust the coupling between the Gaussians, ensuring they capture multiple events in a query without excessive overlap. This approach allows for better representation of complex temporal locations in videos.

Weakly supervised methods typically encounter challenges owing to incomplete annotations and limitations in language queries, including constrained vocabulary and restricted sentence structures. OmniD [2] addresses these challenges by leveraging a Large Language Model (LLM) to paraphrase sentence queries, providing richer training signals without additional annotations. Through a bootstrapping framework, OmniD trains a sophisticated teacher model composed of sub-models, which provide pseudo-labels for temporal boundaries. The method then uses semantic equivariance distillation to transfer this knowledge to a smaller student model, improving performance in weakly supervised settings.

While most weakly supervised grounding models rely on the RGB frameworks, MMDist utilizes multi-modal inputs, such as optical flow and audio, to capture more comprehensive video features. Although this multi-modal data integration increases grounding accuracy, it also adds computational costs. To mitigate this, MMDist [3] trains a multi-modal teacher model and distills its knowledge into a single-modal student model, which only requires RGB frames for inference. This approach enhances model robustness and accuracy without significantly increasing computational overhead during inference.

2.1.3 Others

In the previous sections, we delve into the two primary methods of tackling *Video Grounding* tasks, i.e., fully supervised and weakly supervised approaches. While these approaches have demonstrated their efficacy in various scenarios, they come with their inherent limitations.

On the one hand, fully supervised methods demand an abundant and expensive resource, i.e., meticulously annotated data. The process of manually labeling events in videos is labor-intensive, limiting their practicality in large-scale applications. This drawback raises a critical question: can we achieve robust *Video Grounding* without relying on copious amounts of costly human annotations?

On the other hand, while weakly supervised methods provide a more cost-effective alternative, they frequently exhibit lower precision and accuracy compared to fully supervised methods. This limitation underscores the need to explore alternative strategies that can bridge the gap between the richness of fully supervised data and the cost-effectiveness of weakly supervised approaches.

In response to these challenges, this section will delve into a spectrum of supervision techniques, beginning with Point-Level Supervision, a nuanced approach that focuses on specific temporal points of interest within videos. We will then explore semi-supervised learning, which seeks to leverage both labeled and unlabeled data to improve video grounding. We will then look at unsupervised approaches, in which models do not use video-query pairs or annotated segments; instead, they learn exclusively from unlabeled data. Lastly, we will talk about zero-shot techniques, which use pre-trained vision-text models to produce predictions for video grounding tasks directly, avoiding task-specific training.

Each of these approaches offers a unique perspective on how we can advance the field of *Video Grounding* while circumventing the limitations associated with traditional supervision methods. By examining these alternatives, we aim to broaden the horizons of video understanding and inspire novel approaches.

2.1.3.1 Point-Level Supervised Methods

Point-Level Supervision is a form of weakly supervised learning that differs from conventional labeling approaches by relying solely on a single frame randomly selected within the target event. This method simplifies the annotation process compared to fully supervised labeling methods.

The ViGA method [27], inspired by weakly supervised learning techniques, segments videos into multiple clips and assigns varying weights to each clip to leverage annotated points effectively. For network training, a Gaussian Alignment Module is introduced. This module generates a Gaussian distribution centered on the annotated timestamps, which is utilized to weight all video frame features when computing the Gaussian Label-Smoothed Noise Contrastive Estimation loss. Additionally, this Gaussian heuristic distribution is employed in the Query Attention Guide Kullback–Leibler Divergence Loss (QAG-KL) to guide the network's learning process. During inference, proposals of varying sizes are sampled around the anchor frame, ensuring that each proposal includes the anchor frame. The proposal with the highest dot-product similarity to the sentence feature vector is selected as the final output.

CFMR [59] comprises three Transformer-based components: (1) the Video Concept Encoder (VCE), (2) the Text Concept Encoder (TCE), and (3) the Semantic Reconstructor (SR). The VCE and TCE modules encode videos and text queries into concept vectors representing diverse semantics, while the SR module uses semantic cues to learn concepts of each modality during training. Temporal anchors are represented as Gaussian distributions along the temporal dimension and are utilized in the VCE to re-weight self-attention matrices, enhancing video content within the corresponding moments in the encoded video features. Contrastive Learning is employed as the loss function, with annotated points serving as the sole positive examples. During testing, the SR module is disabled, and cosine similarity is computed among video and text concepts for moment retrieval. This approach bypasses online cross-modal interactions, greatly improving efficiency and conserving computational resources.

2.1.3.2 Semi-supervised Methods

Semi-supervised Learning (SSL) represents a prominent class of techniques within machine learning, tailored for acquiring knowledge from a limited pool of labeled data in a vast reservoir of unlabeled data. These approaches can be broadly classified into two core categories: pseudo label generation and consistency regularization. Pseudo label generation involves the creation of synthetic labels for unlabeled data using a teacher model, which is subsequently employed to refine the learning of the student model. In contrast, consistency regularization methods encourage models to produce consistent predictions when presented with slightly perturbed input data.

In the context of video grounding models, the S^4TLG [87] model adopts a dual-phase training strategy. Initially, a teacher model is pre-trained using a restricted set of annotated data in a supervised manner. Subsequently, in the second phase, the teacher model generates pseudo labels for unlabeled data at each epoch. To enhance the learning process during semi-supervised training, sequential perturbations are introduced to augment the input data for both teacher and student models. Additionally, a self-supervised learning framework is employed, incorporating intra-modal and inter-modal contrastive learning to foster video feature representations that align with video content consistency and video-text alignment. These sequential perturbations serve to create diverse perspectives of the input videos, thereby compelling the model to discern underlying semantic features while remaining invariant to these perturbations.

In the case of the SVPTR [57] model, semi-supervised learning is applied to the domain of Video Paragraph Grounding (VPG). In VPG, the task involves localizing multiple relevant moments in an untrimmed video based on a descriptive paragraph, rather than a single sentence. To enhance the model's robustness and generalization, feature perturbations are administered to both video and text modalities. Video perturbations entail the introduction of random temporal shifts, while text perturbations encompass the random masking of portions of text within the paragraph. Moreover, a dropout strategy is applied to both modalities. During training, the student model receives input data from the feature perturbation module and predicts timestamps, whereas the teacher model is fed with the original, unaltered data to generate predictions. A consistency loss is computed based on the predictions of both models, facilitating joint learning from labeled and unlabeled data. This approach is particularly effective for training the model to perform Video Paragraph Grounding in a semi-supervised manner.

2.1.3.3 Unsupervised Methods

Unsupervised methods in video grounding aim to develop models without relying on paired video-query knowledge or annotated video segments. These approaches innovate by using external knowledge and creative techniques to transfer information from related domains or generate pseudo-supervision.

Gao and Xu [35] pioneer the development of unpaired video moment retrieval models, avoiding the use of labeled video-sentence pairs. They leverage pre-trained image-sentence embedding models and visual concept detectors to bridge the gap between image and video domains. The model consists of two main components: a video-conditioned sentence generator and a Graph Neural Network (GNN)-based moment localizer. The sentence generator is trained to create sentences that accurately depict visual concepts found in the video, such as scenes, objects, and actions. Simultaneously, the GNN-based localizer selects relevant video clips by exploiting temporal relations between video segments. By combining these elements, the model generates sentence and moment representations and matches them using a pre-trained image-sentence embedding space, enabling efficient moment retrieval.

DSCNet [81] mines deep semantic features from query sets and leverages an encoder-decoder model to encode queries and cluster them into semantic centers. DSCNet also incorporates a video-based semantic aggregation module, consisting of a specific attention branch and a foreground attention branch, to compose activities within videos guided by these deep semantic features. This approach aims to generate more accurate video representations and refine the model through iterative learning using pseudo labels derived from deep semantic features.

PSVL [99] introduces a pseudo-supervision strategy, which involves generating pseudo-temporal regions and corresponding sentences to train video grounding models. This pseudo-supervision serves multiple purposes: it provides interpretable resources for model training, reduces human annotation costs, and can be applied to fully supervised models. The model faces two challenges: finding meaningful temporal regions and creating corresponding query sentences. To overcome these, visual information is clustered to generate candidate regions, while nouns and verbs are extracted using object detectors and noun-verb statistical co-occurrence patterns. These pseudo queries are then used to train a simple video grounding model, effectively bridging the gap between unannotated and supervised settings.

PZVMR [133] extends the idea of pseudo-supervision by incorporating prompt learning. Without annotated sentences or timestamps, the model generates pseudo-supervision through temporal proposal generation and pseudo-query construction. The framework uses pre-trained CLIP models to recognize potential moment proposals and match them with pseudo-queries. The model leverages prompt learning to refine moment-query pairs, using visual context features and verb predictions generated through object detection and prompt tuning. This approach effectively distills knowledge from pre-trained models, outperforming unsupervised and even weakly supervised methods in video moment retrieval tasks.

Kim et al. [62] propose a novel language-free framework, eliminating the need for generating pseudo-sentences from visual features. Instead of creating pseudo textual queries from visual information, they utilize the image-language pre-training model of CLIP to directly align visual features with language semantics. Temporal proposals are generated from untrimmed videos, and a dominant visual feature is selected as the pseudo textual representation. This approach avoids the time-consuming task of generating textual embeddings and streamlines the learning process by directly using visual features for grounding, leading to efficient and effective unsupervised video grounding.

LoVe [85] introduces a new approach by treating pseudo-query generation as a video-to-concept retrieval task. It uses a broad open-concept pool, built from text corpora, and the CLIP model to retrieve relevant concepts from video inputs. In the verification stage, irrelevant concepts are discarded, leaving only the most accurate for use as pseudo-queries. LoVe also employs a Temporal Context Encoding (TCE) module to capture long-term dependencies, improving the accuracy of pseudo-queries and enhancing video grounding performance.

2.1.3.4 Zero-Shot Methods

Supervised methods for video grounding necessitate extensive labeled datasets, comprising numerous pairs of action moments and corresponding sentences. Although weakly supervised and unsupervised approaches aim to mitigate the cost associated with data annotation, they still require some level of data collection. Furthermore, all methodologies—whether supervised, weakly supervised, or unsupervised—require training, and their performance is significantly influenced by the distribution of the training data. As a result, these methods often struggle to generalize across different scenarios, especially in out-of-domain contexts. Considering the inherent complexity and variability of real-world videos, the efficacy of current video grounding techniques remains constrained. These challenges pertaining to data requirements and generalization capabilities have prompted researchers to investigate training-free approaches, utilizing pre-trained models to facilitate zero-shot video grounding.

AZVG [134] adopts a pre-trained image-text model, such as CLIP, to obtain text features from the input sentence and image features from video frames. By comparing the cosine similarity between the sentence and the frames, the method identifies frames most related to the sentence, which serve as temporal anchors. The video grounding task is then decomposed into three sub-tasks: identifying anchor frames, segmenting the video into atomic actions, and merging adjacent regions to form final proposals. Once the text and frame features are acquired, the CAG (Candidate Anchor Generation) module is used to create anchor candidates. These anchors are further refined by the ATPD (Atom-based Time Period Detection) module, which expands the anchors into time periods. The result is a set of top-n candidate time periods that match the given sentence.

ZSTEL [119] utilizes large-scale pre-trained vision-language models, leveraging their advanced image and text encoders to identify the most pertinent video frames. A query-guided optimization mechanism is implemented to pinpoint local frame regions where events are most likely to occur, thereby enhancing overall event localization accuracy. Additionally, ZSTEL introduces a zero-shot event proposal generation approach that predicts the temporal boundaries of events. The model's performance is further improved through an efficient greedy event sampling strategy, which enables the generation of multiple high-quality proposals and significantly boosts top-5 accuracy.

Diwan et al. [29] employ a three-step pipeline for zero-shot video grounding. First, it generates moment proposals using an off-the-shelf shot transition detector. Next, it matches moment queries with video content using models like CLIP and UniVL, designed for video captioning. The final step is a post-processing phase that applies a simplified Watershed algorithm to refine the results.

Wattasseril et al. [138] propose a zero-shot approach that extracts frame-level features using off-the-shelf Vision-Language Pre-trained models. Sparse frame sampling is employed to enhance efficiency, while BLIP-based models are used to compute frame and query embeddings. These models, known for their strong performance on vision-language tasks, are repurposed for video moment retrieval by matching video frames and query text, followed by simple merging strategies to finalize the temporal localization.

Luo et al. [88] utilize large-scale pre-trained video-text models to align moment-level features with textual queries. The key challenge is to adapt these models, originally designed for video-level retrieval, to the finer granularity required for moment-level localization. They accomplish this by breaking up the video into short segments, honing the features of the segments using query-specific data, and then using a bottom-up approach to produce proposals. These proposals are based on combinations of refined snippet features that best align with the query, ensuring accurate moment boundaries despite the absence of labeled training data.

VTG-GPT [148] employs Baichuan2 to rephrase the original query, removing biases introduced by human annotation. Furthermore, MiniGPT-v2 is employed to convert visual content into textual descriptions, thereby enhancing the linguistic processing of video data. Temporal proposals are generated by integrating the debiased query with video captions, enabling precise and efficient localization.

TFVTG [177] prompts large language models (LLMs) to analyze queries containing multiple events, breaking them down into sub-queries. Each sub-query corresponds to a single event, and pre-trained vision-language models (VLMs) are used to locate the relevant video segments. TFVTG tackles the challenge of dynamic transitions in videos by explicitly modeling both the transition phase and the ensuing static state. Through assessing the similarity between dynamic and static segments, the method identifies the top proposals for each sub-query. These findings are subsequently integrated, taking into account the event sequence and interrelationships, to generate the final localization results.

Grounding-Prompter [25] reformulates the *Video Grounding* task as a long-textual task and aims to enhance large language models with the ability to conduct complex temporal reasoning in long videos. Specifically, this method leverages both speech and visual information to prompt Large Language Models (LLMs) for the Temporal Scene Graph (TSG) task. Initially, the speech is transcribed, and captions are generated for sparsely sampled frames. This process aligns auditory and visual data with temporal information, resulting in a concise textual representation of the task. To enhance temporal reasoning, a Boundary-Perceptive Prompting strategy is introduced, comprising three key components: a multiscale denoising Chain-of-Thought (CoT) mechanism that integrates global and local semantics while progressively filtering out noise, validity principles to ensure LLMs produce predictions in appropriate formats, and one-shot In-Context Learning (ICL) to improve the LLM's understanding of TSG and temporal reasoning through imitation.

LLM4VG [33] is a comprehensive benchmark designed to systematically evaluate the performance of large language models on video grounding tasks. It employs two strategies to achieve video grounding. The first strategy leverages video Large Language Models (LLMs) trained on text-video datasets, which directly accept video content and grounding task instructions as input to generate predictions. The second strategy incorporates pre-trained visual description models to transform video content into textual descriptions, thereby bridging the gap between visual and textual information. In this approach, prompts are meticulously designed to integrate video grounding instructions with visual description information generated by various models, including caption-based generators for direct descriptions and Visual Question Answering (VQA)-based models to enhance the granularity of the descriptions. This method underscores the capability of LLMs to understand the temporal dimensions of video grounding tasks by compensating for the limitations of caption models in incorporating keywords from the grounding query.

2.2 Datasets and Evaluations

After discussing the methodologies for *Video Grounding*, ranging from supervised learning settings to unsupervised learning setting, we next place our focus on datasets and evaluations.

2.2.1 Benchmark Datasets

In this section, we present benchmark datasets and evaluation metrics for *Video Grounding*, and provide detailed performance comparisons among the above mentioned approaches.

Several datasets for *Video Grounding* from different scenarios with their distinct characteristics have been proposed in the past few years. There is no doubt that the effort of creating these datasets and designing corresponding evaluation metrics do promote the development of *Video Grounding*. Tables 2.1 and 2.2 provide an overview about the statistics of public datasets. Table 2.1 gives an overall introduction about the videos and annotated query-moment pairs. As we can see that some datasets (i.e., TACoS, Charades-STA) are constrained in a narrow and specific scene (e.g., kitchen or indoor activity), while others (i.e., DiDeMo, ActivityNet Captions) involve more complicated activities in open domains. The number of queries would be equal to the number of samples (query-moment pairs) because

Table 2.1 The statistics of videos and annotations of the benchmark datasets

	Video statistics				Annotation statistics			
	# Videos	Aver. video duration (s)	Domain	Video source	# Queries	# Moments	Aver. moment duration (s)	Aver. query length
DiDeMo	10,642	29.3	Open	Flickr	41,206	28,925	6.9	7.5
TACoS	127	286.6	Cooking	Lab Kitchen	18,227	7069	27.9	9.4
Charades-STA	9848	30.6	Indoor activity	Activity	16,124	11,767	8.1	6.2
ActivityNet Captions	14,926	117.6	Open	Activity	71,957	71,718	37.1	14.4

Table 2.2 Language statistics of the benchmark datasets

	Vocabulary statistics (Number of used unique tokens			Sentence statistics (Aver. number per query)		
	Verb	Adjectives	Nouns	Verbs	Adjectives	Nouns
DiDeMo	1.50K	1.40K	4.30K	1.20	0.58	2.64
TACoS	0.58K	0.42K	0.98K	1.48	0.23	2.64
Charades-STA	0.25K	0.17K	0.63K	1.26	0.06	2.40
ActivityNet Captions	2.60K	2.90K	8.90K	2.56	0.66	3.73

each query refers to exactly one moment, but multiple queries may refer to the same moment (a moment in this context refers to a video segment that can be identified by a {video id, start timestamp, end timestamp} triplet). In contrast, the number of moments would be less than that, which is actually the number of unique {video id, start timestamp, end timestamp} triplets. Moreover, the detailed language statistics are reported in Table 2.2. Larger vocabulary size and average number of verbs/adjectives/nouns tokens indicate greater challenges in textual semantic understanding. Obviously, the sentences in ActivityNet Captions are the most difficult and those of Charades-STA are relatively simple with the smallest action (verb) set. We will introduce these four datasets more concretely as follows.

DiDeMo [45] This dataset is collected from Flickr, and consists of various human activities uploaded by personal users. Hendricks et al. [45] split and label video segments from original untrimmed videos by aggregating 5-s clip units, which means the lengths of ground truth segments are times of 5 s. They claim that this trick is for avoiding ambiguity in labeling and accelerating the validation process. However, such a length-fixed issue makes the retrieval task easier, since it compresses the searching space into a set with limited candidates. The data split is also provided by [45], with 33,005/4180/4021 video-sentence pairs for training/validation/test, respectively. Besides, a new dataset TEMPO [46] involving more temporally-related events is collected based on the DiDeMo, which is explored by some works as well [117, 161].

TACoS [105] TACoS is built on MPII-Compositive dataset [109]. It contains 127 complex videos featuring cooking activities, and each video has several segments annotated by sentence descriptions illustrating people's cooking actions. The average length of videos in TACoS is around 300 s, which is much longer than that of other benchmark datasets. The total amount of query-moment pairs is 18,227 in this dataset, and 50%, 25%, 25% of which are used for training, validation, and testing, respectively.

Charades-STA [36] Charades-STA is built upon Charades [112], which is originally collected for video activity recognition, and consists of 9848 videos depicting human daily indoor activities. Specifically, Charades contains 157 activity categories and 27,847 video-level sentence descriptions. Based on Charades, Gao et al. [36] construct Charades-STA with a semi-automatic pipeline, which parses the activity label out of the video description first and aligns the description with the original label-indicated temporal intervals. As such, the yielded (description, interval) pairs can be seen as the (sentence query, target segment) pairs for *Video Grounding*. Since the length of the original description in Charades-STA is quite short, Gao et al. [36] further enhances the complexity of the description by combining consecutive descriptions into a more complex sentence for testing.

ActivityNet Captions [66] ActivityNet Captions is originally proposed for dense video captioning upon ActivityNet dataset [6], and the query-moment pairs in this dataset can naturally be utilized for *Video Grounding*. With the greatest number of videos, ActivityNet Captions aligns videos with a sequence of temporally annotated sentence descriptions. On average, each of the 20k videos contains 3.65 temporally localized sentences, resulting in a total of 100k sentences. Each sentence has an average length of 13.48 words. The sentence length is also normally distributed. Since the official test set is withheld for competitions, most *Video Grounding* works merge the two available validation subsets "val1" and "val2" as the test set. In summary, there are 10,009 videos and 37,421 query-moment pairs in the training set, and 4917 videos and 34,536 query-moment pairs in the test set.

The four datasets introduced above are the most commonly used ones in *Video Grounding* tasks. However, several other datasets have been developed to extend this task in various ways. The following section covers a few datasets that expand on the fundamental *Video Grounding* framework.

TVR [68] is designed for the Video Corpus Grounding task, where the model must first identify the relevant video from a large video corpus containing the query segment and then perform temporal grounding within that video. TVR includes 109,000 queries and 21,793 videos from six TV series, including crime, medical drama, and sitcoms. Each video is approximately 76.2 s long on average. Each query is labeled based on the modalities required for inference, and the dataset includes corresponding subtitles. Of the queries, 74.2% can be answered with only video information, 9.1% with only subtitle information, and 16.6% with both video and subtitle information. Compared to existing datasets, TVR queries typically involve more people and actions. For example, "Chandler points at Joey while describing a woman who wants to date him." TVR is split into 80% for training, 10% for validation, 5% for public testing, and 5% for private testing. The public test set is used for leaderboard evaluation, while the private test set is reserved for future challenges.

QVHighlights [69] can be used for both *Video Grounding* and highlight detection tasks. For the *Video Grounding* task, one or more segments within a video may match the given query, with an average of 1.8 segments per query. The queries are manually annotated, and each matched segment is further broken down into multiple 2-s sub-segments. Salience scores for each sub-segment range from 1 to 5, allowing for more in-depth understanding of the connection between sentence queries and visual features. QVHighlights contains 10,148 videos, each 150 s long, sourced from YouTube, covering categories such as daily activities, travel logs, and news. The dataset is split into 70% for training, 15% for validation, and 15% for testing.

MAD [114] is a long video grounding dataset collected from movies. The average movie length is 110.8 minutes, while the average segment length is only 4.1 s, making it extremely challenging to locate such short segments within long videos. To build this large-scale dataset, MAD utilizes audio descriptions intended for visually impaired audiences. Specifically, these descriptions preserve the movie's dialogue and add a narrator to describe the visual content. By converting these audio descriptions into text, the dataset automatically creates scalable annotations for video segments. The MAD dataset contains 488 movies for training, 50 for validation, and 112 for testing, with 280,183, 32,064, and 72,044 queries in each respective set.

Ego4D-NLQ [41] is part of the Ego4D dataset, which consists of first-person videos collected from wearable cameras. Video collectors perform everyday activities while wearing the camera, creating a diverse dataset of egocentric videos. Among the many tasks designed for Ego4D, the Natural Language Queries (NLQ) task is the most relevant to *Video Grounding*. In this task, queries are phrased as natural language questions, such as "What did I put in the drawer?" The goal is to locate the time segments in the video that answer the query. The Ego4D-NLQ dataset includes 998 videos for training, 328 for validation, and 333 for testing, with 12,991, 3874, and 4005 queries, respectively. Each video is approximately 8 minutes long, and the average segment length is 8.3 s.

2.2.2 Evaluation Metrics

There are two types of metrics for *Video Grounding*, i.e., R@n,IoU=m and mIoU, both of which are first introduced for *Video Grounding* in [36]. Since IoU (Intersection over Union) is widely used in object detection to measure the similarity between two bounding boxes, similarly for *Video Grounding*, as illustrated in Fig. 2.14, many *Video Grounding* methods adopt temporal IoU to measure the similarity between the ground truth moment and the predicted one. The ratio of intersection area over union area ranges from 0 to 1, and it will be equal to 1 when these two moments are totally overlapped.

Thereby, one of the metrics is mIoU (i.e., mean IoU), a simple way to evaluate the results through averaging temporal IoUs of all samples. The other commonly-used metric is R@n, IoU=m [50]. When one segment from the top n retrieved

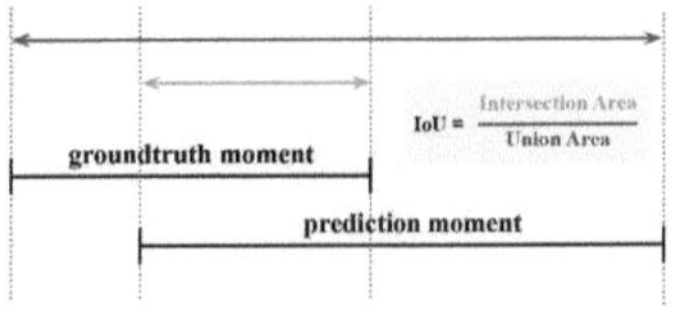

Fig. 2.14 The illustration of Temporal IoU (Intersection over Union)

segments has a temporal IoU with the ground truth segment greater than m, sample i is considered positive. This can be expressed as $r(n, m, q_i) = 1$. Otherwise, $r(n, m, q_i) = 0$. R@n, IoU=m is the percentage of positive samples over all samples (i.e., $\frac{1}{N_q} \sum_i r(n, m, q_i)$).

The community is accustomed to setting $n \in \{1, 5, 10\}$ and $m \in \{0.3, 0.5, 0.7\}$. Usually, $n = 1$ when the method adopts a proposal-free manner (i.e., belongs to either anchor-free or RL-based frameworks). Moreover, it is worth noting that MCN [45] adopts a particular metric with the IoU threshold $m = 1.0$ since the ground truth segments in DiDeMo are generated by aggregating the clip units of 5 s, and MCN also employs a matching-based method thus the predicted moment has a chance to fully coincide with the target moment, satisfying such a high IoU threshold.

2.2.3 *Performance Comparisons*

In this section, we give a thorough performance comparison of the aforementioned approaches based on four benchmark datasets. For convenience and fairness, we uniformly adopt $n = 1$ and $m \in \{0.3, 0.5, 0.7\}$ for the metric of R@n,IoU=m. We consistently report the best results for each method as documented in the literature, despite the possibility that different kinds of extracted visual features may affect the grounding accuracy. Table 2.3 lists all the experimental results grouped by their categories (i.e., belonging to two-stage, single-stage, RL-based or weakly supervised methods) which are segmented by double horizontal lines. Table 2.6 separately reports the experimental results on the DiDeMo dataset with metrics of R@$\{1, 5\}$,IoU=1.0 and mIoU.

Two-Stage Method

As shown in Tables 2.3 and 2.4, the overall performance of two-stage methods seems poorer than other approaches. The potential reasons can be categorized into three aspects: (1) Many two-stage methods coarsely integrate video and sentence features, thereby neglecting the fine-grained interactions between visual and textual elements necessary for precise video grounding. (2) Separating the generation of candidate segments from the query-moment matching process prevents global optimization of the model, which in turn affects overall performance. (3) Establishing matching relationships between sentence queries and individual segments isolates local video content from the global video context, potentially compromising temporal grounding accuracy.

Specifically, for the sliding window (SW)-based methods, all of them achieve the lowest grounding accuracy on the TACoS compared to the other three datasets with the same metrics. The reason is that the cooking activities in TACoS take place in the same kitchen scene with only some slightly varied cooking objects (e.g., chopping board, knife, and bread). Thus, it is hard to do temporal location predictions for

Table 2.3 The performance comparison of fully supervised *Video Grounding* methods grouped by their categories (SW: sliding window-based, PG: proposal-generated, AB: anchor-based, AF: anchor-free, OT: other single-stage methods, RL: reinforcement learning based)

		DiDeMo			TACoS			Charades-STA			ActivityNet captions		
Type	Method	0.3	0.5	0.7	0.3	0.5	0.7	0.3	0.5	0.7	0.3	0.5	0.7
SW	MCN [45]	–	–	–	–	–	–	13.57	4.05	–	–	–	–
	CTRL [36]	–	–	–	18.32	13.3	–	–	21.42	7.15	–	–	–
	MCF [139]	–	–	–	18.64	12.53	–	–	–	–	–	–	–
	ROLE [77]	29.4	15.68	–	–	–	–	25.26	12.12	–	–	–	–
	ACRN [76]	–	–	–	19.52	14.62	–	–	–	–	–	–	–
	SLTA [55]	–	30.92	17.16	17.07	11.92	–	38.96	22.81	8.25	–	–	–
	VAL [115]	–	–	–	19.76	14.74	–	–	23.12	9.16	–	–	–
	ACL-K [38]	–	–	–	24.17	20.01	–	–	30.48	12.2	–	–	–
	MMRG [159]	–	–	–	57.83	39.28	–	71.6	44.25	–	–	–	–
PG	QSPN [144]	–	–	–	–	–	–	54.7	35.6	15.8	45.3	27.7	13.6
	SAP [13]	–	–	–	–	18.24	–	–	27.42	13.36	–	–	–
AB	TGN [17]	–	–	–	21.77	18.9	–	–	–	–	45.51	28.47	–
	MAN [160]	–	–	–	–	–	–	–	46.53	22.72	–	–	–
	CMIN [162]	–		–	24.64	18.05	–	–	–	–	63.61	43.4	23.88
	SCDM [155]	–	–	–	26.11	21.17	–	–	54.44	33.43	54.8	36.75	19.86
	CBP [129]	–	–	–	27.31	24.79	19.1	–	36.8	18.87	54.3	35.76	17.8
	2D-TAN [165]	–	–	–	37.29	25.32	–	–	39.7	23.31	59.45	44.51	26.54
	FVMR [34]	–	–	–	41.48	29.12	–	–	55.01	33.74	60.63	45	26.85
	SMRN [12]	–	–	–	42.49	32.07	–	–	43.58	25.22	–	42.97	26.79
	RMN [78]	–	–	–	32.21	25.61	–	–	59.13	36.98	67.01	47.41	27.21
	FIAN [104]	–	–	–	33.87	28.58	–	–	58.55	37.72	64.1	47.9	29.81
	CSMGAN [79]	–	–	–	33.9	27.09	–	–	–	–	68.52	49.11	29.15
	SMIN [130]	–	–	–	48.01	35.24	–	–	64.06	40.75	–	48.46	30.34
	Zhang et al. [171]	–	–	–	48.79	37.57	–	–	–	–	–	48.02	31.78
	VLG-Net [113]	25.57	71.65	–	45.46	34.19	–	–	–	–	–	46.32	29.82

AF	ABLR [156]	–	–	–	18.9	9.3	–	–	–	–	55.67	36.79	–
	DEBUG [84]	–	–	–	23.45	–	–	54.95	37.39	17.69	55.91	39.72	–
	GDP [20]	–	–	–	24.14	–	–	54.54	39.47	18.49	56.17	39.27	–
	PMI [21]	–	–	–		–	–	55.48	39.73	19.27	59.69	38.28	17.83
	ExCL [39]	–	–	–	44.4	27.8	14.6	61.4	41.2	21.3	62.1	41.6	23.9
	DRN [158]	–	–	–	–	23.17	–	–	45.4	26.4	–	42.49	22.25
	HVTG [14]	–	–	–	–	–	–	61.37	47.27	23.3	57.6	40.15	18.27
	TMLGA [107]	–	–	–	24.54	21.65	16.46	67.53	52.02	33.74	51.28	33.04	19.26
	LGI [97]	–	–	–	–	–	–	72.96	59.46	35.48	58.52	41.51	23.07
	VSLNet [164]	–	–	–	29.61	24.27	20.03	70.46	54.19	35.22	63.16	43.22	26.16
	CPN [174]	–	–	–	47.69	36.33	21.58	75.53	59.77	36.67	62.81	45.1	28.1
	DORi [108]	–	–	–	31.8	28.69	24.91	72.72	59.65	40.56	57.89	41.35	26.41
	CI-MHA [154]	–	–	–	–	–	–	69.87	54.68	35.27	61.49	43.97	25.13
	PEARL [172]	–	–	–	42.94	32.07	18.37	71.9	53.5	35.4	–	–	–
OT	BPNet [142]	–	–	–	25.96	20.96	14.08	55.46	38.25	20.51	58.98	42.07	24.69
	DPIN [128]	–	–	–	46.74	32.92	–	–	47.98	26.96	62.4	47.27	28.31
	CBLN [80]	–	–	–	38.98	27.65	–	–	61.13	38.22	66.34	48.12	27.6
RL	R-W-M [44]	–	–	–	–	–	–	–	36.7	–	–	36.9	–
	SM-RL [127]	–	–	–	20.25	15.95	–	–	24.36	11.17	–	–	–
	TripNet [42]	–	–	–	–	–	–	51.33	36.61	14.5	48.42	32.19	13.93
	TSP-PRL [141]	–	–	–	–	–	–	–	45.45	24.75	56.02	38.82	–
	STRONG [7]	–	–	–	72.14	49.73	18.29	78.1	50.14	19.3	–	–	–
	AVMR [8]	–	–	–	72.16	49.13	–	77.72	54.59	–	–	–	–

Table 2.4 Performance comparison of DETR series methods. For video features, SlowFast and CLIP-ViT-B/32 are used, and text features are extracted using CLIP. "w/InternVideo2-6B" denotes the use of InternVideo2-6B [136] features. "w/PT" refers to weakly supervised pre-training with ASR captions

Method	QVHighlights		Charades-STA		TACoS	
	0.5	0.7	0.5	0.7	0.5	0.7
Moment-DETR [69]	52.89	33.02	53.63	31.37	24.67	11.97
Moment-DETR w/PT	59.78	40.33	55.65	34.17	–	–
UMT[a] [82]	56.23	41.18	–	–	–	–
EaTR [54]	–	–	55.20	33.10	–	–
BM-DETR [61]	60.12	43.05	59.48	38.33	35.42	–
QD-DETR [95]	62.40	44.98	57.31	32.55	36.77	21.07
UVCOM [143]	63.55	47.47	59.25	36.64	–	–
CG-DETR [94]	65.43	48.38	58.44	36.34	39.61	22.23
CG-DETR w/InternVideo2-6B	71.42	56.45	70.03	48.95	–	–
BAM-DETR [67]	62.71	48.64	59.95	39.38	41.54	26.77
TR-DETR [120]	64.66	48.96	57.61	33.52	–	–
RGTR [121]	65.50	49.22	57.93	35.16	40.31	24.32

[a] UMT additionally leverages audio features

such fine-grained activities. Meanwhile, the lengths of videos in TACoS are also longer, which will greatly increase the target segment searching space and bring more difficulties. Obviously, MMRG outperform other SW-based methods with great gains on both TACoS and Charades-STA. Despite using the same moment sampling strategies with CTRL, the multi-modal relational graph MMRG employs can capture subtle differences of candidate moments from the same video and the customized self-supervised pre-training tasks further improve the visual features. Regardless of MMRG, ACL-K also significantly outperforms the remaining SW-based methods on TACoS and Charades-STA, proving the effectiveness of aligning the activity concepts mined from both textual and visual parts. MCN gets the most inferior results on the Charades-STA, which shows that its simple multi-modal matching and ranking strategy for candidate segments cannot deal well with the segments of various and flexible locations. However, CTRL, ACRN, ROLE, SLTA, VAL, ACL-K and MMRG can adjust the candidate segment boundaries based on the model location offset prediction, which can therefore improve the performance. All of the sliding window-based methods have not conducted experiments on the large-scale ActivityNet Captions dataset, which may be due to the costly computation for multi-scale sliding window sampling.

The proposal-generated (PG) methods achieve superior performance compared to sliding window (SW)-based methods, despite a reduction in the number of proposal candidates. Specifically, QSPN, which incorporates a query-guided segment proposal network and auxiliary captioning loss, significantly outperforms other two-stage methods (with the exception of MMRG) on the Charades-STA dataset. This indicates that the query-guided proposal network can provide more effective candidate moments with finer temporal granularity, thereby eliminating the need to process redundant sliding window sampled moments. Additionally, QSPN has been tested on the ActivityNet Captions dataset, which features richer scenes, and it achieves competitive results comparable to single-stage anchor-based methods. This further validates the effectiveness of captioning supervision and query-guided proposals. Given that videos in the Charades-STA dataset are generally shorter and contain fewer diverse activities, higher emphasis should be placed on metrics with higher IoU thresholds. SAP consistently outperforms SW-based methods on Charades-STA under higher IoU thresholds, attributed to its discriminative generated proposals and additional refinement process.

Single-Stage Method For anchor-based (AB) methods, TGN achieves the lowest performance on TACoS and ActivityNet Captions. CMIN also performs poorly on TACoS. The common inferior accuracy achieved by TGN, CMIN and CBP may attribute to their single-stream anchor-based localization frameworks. With sequential RNNs, they fail to reason complex cross-modal relations on datasets (i.e., TACoS and ActivityNet Captions) of longer video lengths. Instead of employing RNN-styled anchors, both SCDM and MAN use convolutional neural networks to better capture fine-grained interactions and diverse video contents of different temporal granularities, which can achieve better performance (e.g., SCDM performs better than TGN/CMIN and TGN/CBP on TACoS and ActivityNet Captions, respectively). To make further improvement, 2D-TAN extends it to 2D feature maps to model the adjacent relations of various candidate moments of multi-anchors. SMIN and the method proposed by Zhang et al. [171] adopt such a similar 2D structure to model the relationships of candidate moments, also achieving superior results out of AB methods on TACoS, Charades-STA and ActivityNet Captions. Specifically, the model presented by Zhang et al. [171] performs the best on TACoS while SMIN has surpassed other methods on Charades-STA, which also proves the effectiveness of 2D moment relationship modeling. Furthermore, CSMGAN, SMIN and Zhang et al. [171] all achieve superior results on ActivityNet Captions. It is noted that although CSMGAN adopts the similar sequential RNN like TGN, it builds a joint graph for modeling the cross-/self-modal relations which can capture the high-order interactions between two modalities effectively.

For anchor-free (AF) methods, the overall performance is slightly worse than that of AB methods especially with the challenging ActivityNet Caption dataset. More specifically, reading comprehension-inspired methods (ExCL, VSLNet, TMLGA, CPN, DORi, CI-MHA and PEARL) outperform other types of anchor-free methods with a significant gap. However, TMLGA achieves the lowest performance with the metrics of R@1,IoU={0.3, 0.5} on ActivityNet Captions. One possible reason is that the subjectivity of annotation is the hardest to model for this challenging dataset. The dense AF methods including DRN, GDP and DEBUG outperform the early sparse regression network ABLR, justifying the importance of increasing the number of positive training samples. However, the additional regression-based methods including PMI, HVTG and LGI achieve superior performance on ActivityNet Captions, which may result from more effective interaction between visual and textual contents. An obvious observation is that DORi achieves the highest grounding accuracy (with IoU=0.7) among all single-stage methods on TACoS and Charades-STA. Since its spatio-temporal graph is able to model more fine-grained object interactions that change over time. It is noted that L-Net has not been included in the table since the original paper [18] did not report the specific experimental values.

Additionally, other single-stage methods (BPNet, DPIN, and CBLN) which can not be grouped into either anchor-based or anchor-free methods achieve comparable results on three datasets (except DiDeMo). Specifically, CBLN achieves superior performance among all single-stage methods on Charades-STA and ActivityNet Captions, which highlights the advantages of combining anchor-based and anchor-free schemes and its special biaffine-based architecture.

RL-based Method Despite the fact that the overall performance of reinforcement learning (RL)-based methods does not match that of traditional single-stage state-of-the-art (SOTA) methods, they introduce innovative approaches to tackle the Video Grounding task. The sequential decision-making process inherent in RL methods also enhances interpretability. Notably, TSP-PRL surpasses TropNet and R-W-M on ActivityNet Captions and Charades-STA datasets, likely due to its tree-structured policy design, which mirrors the coarse-to-fine human decision-making process. STRONG and AVMR outperform other RL-based frameworks on TACoS, attributed to the efficiency of spatial RL for scene tracking and the incorporation of adversarial learning, respectively. Conversely, R-W-M, TripNet, and SM-RL exhibit relatively lower performance. Specifically, SM-RL achieves the lowest performance on Charades-STA and TACoS, while TripNet performs the worst on ActivityNet Captions.

DETR Series Method As shown in Table 2.4 , the performance of DETR series models has been rapidly improving. The end-to-end optimization characteristic of these models reduces the need for dataset-specific hyperparameters, leading to consistently strong performance across different datasets. Additionally, the table highlights two further strategies for enhancing performance: (1) fine-tuning the model on a large amount of video data before training on the target dataset, and (2) using more powerful visual encoders to extract video features, such as InternVideo2 [136].

Non-fully Supervised Method Due to the significant challenge posed by the unavailability of temporal annotations for ground truth moments during the training phase for weakly supervised (WS) methods, the experimental results on Charades-STA and ActivityNet Captions datasets are notably inferior to those obtained using fully supervised approaches (Table 2.5). It is difficult to conclusively determine which framework—whether multiple instance learning (MIL)-based or reconstruction-based—has a definitive advantage based on overall performance metrics. However, among all WS methods, CCL, VCA, and CRM exhibit superior performance on both Charades-STA and ActivityNet Captions. Their results are also competitive when compared with fully supervised methods. To explore the reasons

Table 2.5 The performance comparison of non-fully supervised *Video Grounding* methods grouped by their categories (WS: weakly supervised, PS: point-level supervised, US: unsupervised, ZS: zero-shot)

		Charades-STA			ActivityNet captions		
Type	Method	0.3	0.5	0.7	0.3	0.5	0.7
WS	WSDEC [31]	–	–	–	41.98	23.34	–
	TGA [93]	32.14	19.94	8.84	–	–	–
	WSLLN [37]	–	–	–	42.8	22.7	–
	EC-SL [15]	–	–	–	44.29	24.16	–
	SCN [72]	42.96	23.58	9.97	47.23	29.22	–
	WSTG et al. [23]	39.8	27.3	12.9	44.3	23.6	–
	VLANet [90]	45.24	31.83	14.17	–	–	–
	FSAN [131]	–	–	–	55.11	29.43	
	MARN [116]	48.55	31.94	14.81	47.01	29.95	–
	RTBPN [166]	60.04	32.36	13.24	49.77	29.63	–
	BAR [140]	44.97	27.04	12.23	49.03	30.73	–
	CCL [167]	–	33.21	15.68	50.12	31.07	–
	VCA [132]	58.58	38.13	19.57	50.45	31	–
	LoGAN [122]	51.67	34.68	14.54	–	–	–
	CRM [51]	53.66	34.76	16.37	55.26	32.19	–
	CNM [175]	60.39	35.43	15.45	55.68	33.33	–
	CPL [176]	66.40	49.24	22.39	55.73	31.37	
	CPI [65]	67.64	50.47	24.38	–	–	–
	CCR [89]	68.59	50.79	23.75	53.21	30.39	–
	UGS [52]	69.16	52.18	23.94	58.07	36.91	–
	SCANet [153]	68.04	50.85	24.07	56.07	31.52	–
	OmniD [2]	68.30	52.31	24.35	57.34	31.60	–
	MMDist [3]	68.90	53.29	25.27	58.69	32.52	–
	PPS [63]	69.06	51.49	26.16	59.29	31.25	–
	IRON [10]	70.71	51.84	25.01	58.95	36.27	–
PS	ViGA [27]	–	45.05	20.27	–	35.79	16.96
	CFMR [59]	–	48.14	22.58	–	36.97	17.33
US	Gao and Xu [35]	46.69	20.14	8.27	46.15	26.38	11.64
	DSCNet [81]	–	–	–	–	–	–
	PSVL [99]	46.47	31.29	14.17	44.74	30.08	14.74
	PZVMR [133]	46.83	33.21	18.51	45.73	31.26	17.84
	Kim et al. [62]	52.95	37.24	19.33	47.61	32.59	15.42
	LoVe [85]	47.74	34.62	20.16	49.26	31.45	15.27
ZS	Luo et al. [88]	56.77	42.93	20.13	48.28	27.90	11.57
	VTG-GPT [148]	59.48	43.68	25.94	47.13	28.25	12.84
	TFVTG [177]	67.04	49.97	24.32	49.34	27.02	13.39

Table 2.6 The evaluation results on DiDeMo (The IoU threshold m = 1.0)

Type	Method	R@1	R@5	mIoU
Fully supervised	TMN [75]	22.92	76.08	35.17
	TGN [17]	24.28	71.43	38.62
	VLG-Net [113]	25.57	71.65	–
	MCN [45]	28.1	78.21	41.08
	MAN [160]	27.02	81.7	41.16
Weakly supervised	TGA [93]	12.19	39.74	24.92
	VLANet [90]	19.32	65.68	25.33
	WSLLN [37]	19.4	53.1	25.4
	FSAN [131]	19.4	57.85	31.92
	RTBPN [166]	20.79	60.26	29.81
	LoGAN [122]	39.2	64.04	38.28

behind this, one key observation is that these methods design specialized training objectives that facilitate better visual-semantic alignment even in the absence of annotated boundary information. Specifically, CCL generates fine-grained supervision signals from counterfactual results for contrastive training. Meanwhile, VCA redefines the Video Grounding problem and introduces a novel loss function for visual co-occurrence alignment learning. Moreover, CRM minimizes mismatches between sentence-moment pairs by expanding the scope to paragraph level, thereby considering the temporal ordering between sentences.

DiDeMo Evaluation Results with Particular Metrics As aforementioned, MCN [45] reports the results on DiDeMo dataset under the IoU threshold m=1.0. Some works [17, 90, 160] following MCN also adopt such metrics to assess their models. We additionally list the evaluation results (i.e., R@$\{1, 5\}$,IoU=1.0 and mIoU) on DiDeMo shown in Table 2.6, which are grouped by the supervision manner. Specifically, LoGAN as a WS method achieves the best performance among both fully and weakly supervised methods, which is also due to the effective visual-semantic representation learning via a latent graph co-attention network. Another observation is that the top-1 recall values for all fully supervised methods are constrained into a certain small range (22–27%). It demonstrates that DiDeMo can not greatly differentiate the performance of methods, which may result from its limitation of taking pre-defined segments as the ground truth.

2.3 Advances on Video Grounding

We discuss several representative advanced topics related to *Video Grounding* in this section, including but not limited to Debiased Video Grounding, Spatio-Temporal Sentence Grounding, Dense Events Grounding, Video-Language Pre-training for

Temporal Grounding, Long-form Video Grounding, Video Corpus Grounding and Identity-Text Video Corpus Grounding.

2.3.1 *Debiased Video Grounding*

Despite the promising results made in *Video Grounding*, there are also some recent works [101, 157] doubting the quality of current datasets and metrics: (1) The joint distributions of starting and ending timestamps of target video segments are strongly biased and almost identical in the training and test splits of current datasets. Without truly modeling the video and sentence data and just fitting such distribution biases in the training set, some baselines can still achieve good results and even outperform some well-designed methods. (2) The annotation of ground truth segment location for *Video Grounding* is ambiguous and subjective and may influence the model evaluation. (3) Current evaluation metrics are easily deceived by the above annotation biases in current datasets and cannot measure the model performance effectively. Since *Video Grounding* is heavily driven by these datasets and evaluation metrics, such problematic benchmarks will influence the research progress of *Video Grounding*, and further mislead this research direction. The limitations of current datasets and evaluation metrics will be discussed in detail below, along with some new approaches to these problems.

Annotation Distribution Biases in Datasets Some recent studies [101, 157] attempt to visualize the temporal location distribution of ground truth segments, finding joint distributions of starting and ending timestamps of ground truth segments identical in training and test sets with obvious distribution biases. For instance, a bias-based method [157] that samples locations from the observed training distribution and uses them as predicted locations of target segments at the inference stage is one of their basic model-free techniques. This bias-based method can achieve good performance even surpassing some well-designed deep models, without any valid visual and textual inputs. Further, Yuan et al. [157] reorganize two benchmark datasets for out-of-distribution testing. They create two different test sets: one test set follows the identical temporal location distribution with the training set, namely test-iid, and the other test set that has quite a different distribution with the training set, namely test-ood. After comparing the experimental results of various baseline methods on these two test sets, they find that for almost all methods, the performance on test-ood drops significantly (c.f., Fig. 2.15), which indicates that existing methods are heavily influenced by temporal annotation biases and do not truly model the semantic matching relationship between videos and texts. Thus, it is crucial for future works to construct debiased datasets and build robust models unaffected by biases. Recently, there have been some attempts to address this issue. For example, Yang et al. [149] designs a causal-inspired framework based on CTRL and 2D-TAN, which attempts to eliminate the spurious correlation between the input and prediction caused by hidden confounder (i.e., the temporal location of moments).

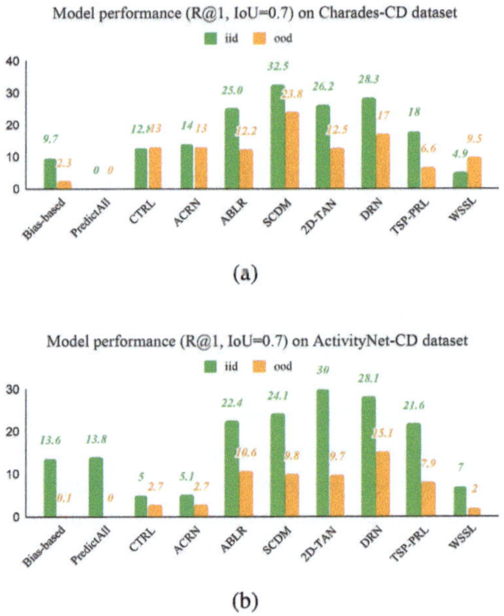

Fig. 2.15 Performance of SOTA *Video Grounding* methods on re-organized data splits. (**a**) Model performance on Charades-CD. (**b**) Model performance on ActivityNet-CD

Moreover, it is worth noting that there are some debiased works [100, 179] that concentrate on other kinds of biases in *Video Grounding* instead of the moment annotation distribution biases. Zhou et al. [179] are devoted to dealing with the biases caused by single-style annotations. The proposed DeNet with a debiasing mechanism can produce diverse yet plausible predictions. Nan et al. [100] propose an approach to approximate the latent confounder set distribution based on the theory of causal inference to deconfound selection biases introduced by datasets (e.g., in datasets, it appears more often that a person is holding a vacuum cleaner than a person is repairing a vacuum cleaner).

Ambiguity of Ground Truth Annotation One recent study [101] also mentions the ambiguous and inconsistent annotations among current *Video Grounding* datasets. Annotating the target segment location of the provided sentence query is a quite subjective task. In certain instances, a single query may be matched with several video segments, or different annotators may arrive at different conclusions regarding the sentence query's grounded location. Therefore, only using one single ground truth to evaluate the temporal grounding results is problematic. Otani et al. [101] suggest to re-annotate the benchmark datasets with multiple ground truth moments for one given sentence query if it exists, as shown in Fig. 2.16, they ask five annotators to re-annotate a video from ActivityNet Captions given the query "a woman is doing somersaults and big jumps alone". These five re-annotated segments corresponding to the query are entirely distinct and do not overlap with the ground truth segment, highlighting the ambiguity and subjectivity inherent in ground truth annotations. Furthermore, they introduce two alternative evaluation metrics that address multiple annotated ground truth moments.

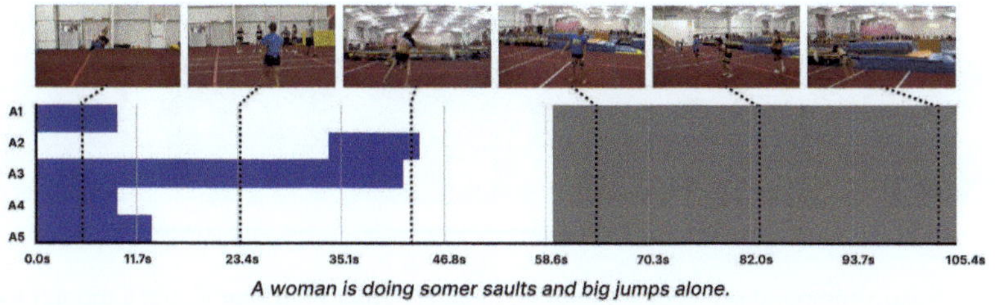

Fig. 2.16 The re-annotation example for ActivityNet Captions. The annotators annotate five different positive segments (shown as blue bars), all of which match the given query. While the original ground truth segment is represented as grey bar

Limitation of Evaluation Metrics Besides the temporal annotation biases in the current dataset, Yuan et al. [157] also find that some characteristics of the datasets may have negative effects on model evaluation. Most of previous *Video Grounding* methods [17, 77, 144, 156, 165] report their scores on some small IoU thresholds like $m \in \{0.1, 0.3, 0.5\}$. However, for ActivityNet Captions, a substantial proportion of ground truth moments are of quite long lengths. Statistically, 40%, 20%, and 10% of sentence queries refer to a moment occupying over 30%, 50%, and 70% of the length of the whole video, respectively. Such annotation biases can obviously increase the chance of correct prediction under small IoU thresholds. Taking an extreme case as an example, if the ground truth moment is the whole video, any predictions with duration longer than 0.3 can achieve R@1,IoU=0.3=1. Thus, the metric R@n, IoU=m with small m is unreliable for current biased annotated datasets. Therefore, to alleviate the above effects, they present a new metric namely discounted-R@n, IoU=m. This new metric considers that the hit score (i.e., $r(n, m, q_i)$) for each positive sample i should not be limited to $\{0, 1\}$. It can be a real number $\in [0, 1]$ depending on the relative distances between the predicted and ground truth boundaries. The formal definition for each sample i is as follows:

$$r(n, m, q_i) = (1 - \text{nDis}(g_i^s, p_i^s)) \times (1 - \text{nDis}(g_i^e, p_i^e)) , \tag{2.16}$$

where the nDis operation calculates the distance between the ground truth and predicted boundaries normalized to [0, 1] by the video length. $(g_i^s, g_i^e)/(p_i^s, p_i^e)$ indicates the (start, end) timestamps of the ground truth/predicted segment for sample i.

2.3.2 *Spatio-Temporal Sentence Grounding*

Spatio-temporal sentence grounding in videos (STSGV, or STVG) is another extension from *Video Grounding* which mainly localizes the referring object/instance as a continuing spatio-temporal tube (i.e., a sequence of bounding boxes) extracted

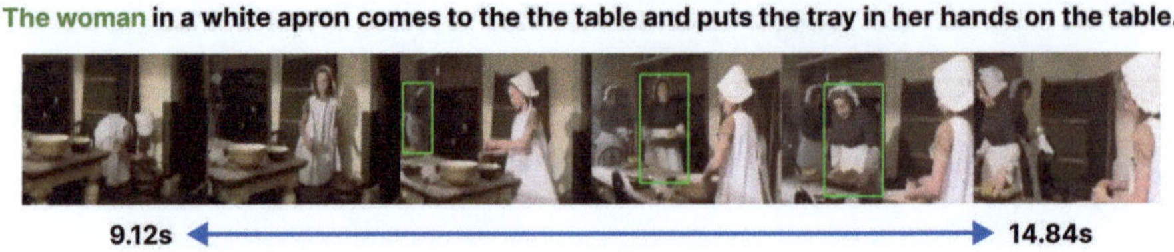

Fig. 2.17 Spatio-temporal sentence grounding, models output both spatial and temporal localization

from an untrimmed video via a natural language description, as shown in Fig. 2.17. Unlike *Video Grounding*, which focuses solely on temporal boundaries, STSGV adds complexity by requiring the localization of both the temporal boundaries and the spatial bounding boxes across frames within the video segment. We will first introduce two commonly-used datasets for STSGV, along with the evaluation metrics. Following this, we will discuss both fully supervised and weakly supervised methods in detail.

Dataset and Evaluation

Vid-STG [169] consists of 99,943 sentence descriptions, which include 44,808 declarative sentences and 55,135 interrogative sentences describing 79 object categories found in 10,303 videos. A declarative sentence makes a statement, such as "A little boy with a Christmas hat is catching a yellow toy". An interrogative sentence asks a question, like "What is the squatting boy on the floor catching". The dataset is split into subsets of 80,684, 8956, and 10,303 distinct sentences and 5436, 602, and 732 distinct videos, for training, validation, and testing sets respectively.

HC-STVG [125] contains videos featuring multiple persons, each annotated with a sentence referring to a specific person. The first version of the dataset, HC-STVG1, is divided into training and test sets, comprising 4500 and 1160 video-sentence pairs, respectively. Each video is 20 s long and is paired with a natural language query, averaging 17.25 words per query. An improved version, HC-STVG2, is split into training and validation sets, with 10,131 and 2000 video-sentence pairs, respectively.

The evaluation metrics for STSGV are based on the calculation of both temporal and visual IoU. For temporal IoU, the m_tIoU metric measures the average temporal Intersection over Union between the selected video clips and the ground truth clips. For visual IoU, we define S_U as the set of frames in either the selected or ground truth clips, and S_I as the set of frames common to both. The vIoU is computed as: $\frac{1}{|S_U|}\sum_{t\in \S_I} IOU(r^t, \hat{r}^t)$, where r^t and $\hat{r}^t$ represent the selected and ground truth regions in frame t, respectively. The m_vIoU metric is the average vIoU across all samples, and vIoU@m metric represents the proportion of samples where vIoU exceeds a specified threshold m.

Fully Supervised Methods

Next, we will introduce fully supervised methods for spatio-temporal Sentence Grounding task. Early models adopt a two-stage approach: first generating object or tube proposals, and then predicting the final tube.

STGRN [169] constructs a spatio-temporal region graph that incorporates both spatial subgraphs within frames and temporal subgraphs across frames. This design allows the model to distinguish subtle differences in object actions, such as "open the door" versus "close the door". By fusing textual clues with the spatio-temporal graph, STGRN applies multi-step cross-modal reasoning to capture complex relationships, such as "a man hugging a baby wearing a red hat". Finally, object tubes are directly retrieved using a spatio-temporal localizer, which first detects temporal boundaries before performing spatial grounding in every frame.

STGVT [125] method jointly leverages multi-modal information from both videos and text. The approach begins by generating tube proposals from bounding boxes linked across consecutive frames. A visual transformer then learns cross-modal interactions between the tube proposals and the query sentence, predicting the relevant tube while trimming irrelevant frames. This method effectively integrates visual and textual data to enhance spatio-temporal grounding accuracy.

OMRN [168] addresses a limitation in STGRN, which does not filter irrelevant regions, potentially hampering effective relation modeling. OMRN focuses on crucial objects mentioned in the sentence, performing fine-grained cross-modal reasoning between these objects and the video regions. Multiple branches are used to model object-aware regions, with the main branch targeting the queried object and auxiliary branches focusing on other objects in the sentence. An object-aware modulation technique enhances relevant region features and suppresses irrelevant ones. A multi-branch relation reasoning mechanism further captures critical object relationships while filtering out irrelevant regions. Finally, a diversity loss ensures each branch focuses on distinct objects, and the spatio-temporal localizer retrieves the target tube by determining temporal boundaries.

However, better representations can be learned by solving both spatial and temporal localization in an end-to-end optimized network. The one-stage approach, which better integrates these tasks, has been used in a number of models.

STVGBert [118] is a one-stage visual-linguistic transformer-based framework designed for spatio-temporal video grounding task. It directly generates spatio-temporal object tubes from input videos and query descriptions without relying on pre-trained object detectors. The method first extracts cross-modal features from a video-text pair and uses these features to produce bounding boxes for each frame while predicting the temporal boundaries. In order to enable more precise spatio-temporal grounding, the model expands upon a visual-linguistic transformer, namely ST-ViLBert, which encodes both spatial and temporal information.

An alternative method is presented by TubeDETR [150], which uses a transformer decoder that independently reasons about temporal and spatial dimensions to perform visual grounding. It leverages time queries to concurrently address temporal localization and visual grounding, while sharing representations across both tasks to generate robust contextualized features. A lightweight branch is utilized to refine temporal details, and the core components of the model effectively encode spatial and multi-modal interactions over sparsely sampled frames. The space-time decoder incorporates self-attention and cross-attention layers to predict object bounding boxes and temporal boundaries.

STCAT [60] addresses the issue of prediction inconsistency in earlier models like STVGBert and TubeDETR, which generate bounding boxes for each frame independently, neglecting temporal consistency. STCAT introduces a video-level multi-modal template to maintain consistency across frames, using a learnable query that guides the grounding process globally. The framework employs a cross-modal spatio-temporal encoder to fuse visual and textual information, dynamically updating the query for each frame during the decoding process. This ensures coherent spatio-temporal tube predictions.

STVGFormer [74] tackles the STVG task with a two-branch architecture: a static branch for spatial grounding and a dynamic branch for temporal grounding. The static branch focuses on identifying the target based on appearance and other static cues, while the dynamic branch handles actions and temporal cues. A static-dynamic interaction block facilitates information exchange between the branches, improving cross-modal understanding by combining complementary information. With each branch having its own prediction head—the dynamic branch forecasting temporal localization, the static branch forecasting spatial grounding—the framework is both efficient and succinct.

Although prior approaches have achieved advancements in STVG, their capability to generalize to out-of-distribution and open-vocabulary samples remains constrained by the limitations inherent in video datasets. VideoGrounding-DINO [137] introduces a novel approach by leveraging the powerful representations of foundation model Grounding-DINO [83], which are trained on extensive image-text data and excel in both closed-set and open-vocabulary scenarios. As shown in Fig. 2.18, this model adopts a DETR-like architecture enhanced with temporal

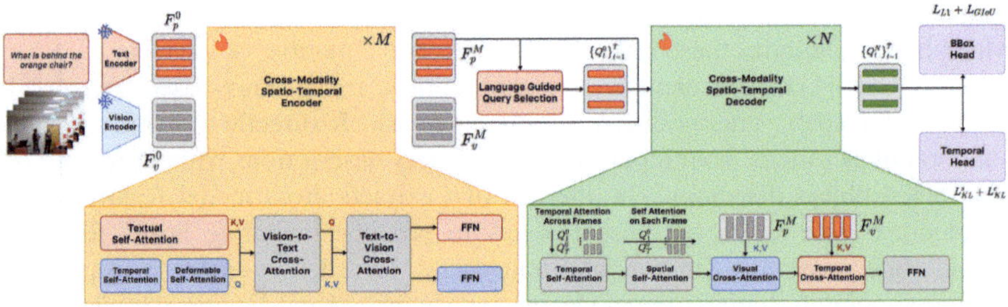

Fig. 2.18 The architecture of VideoGrounding-DINO, which consists of vision and text encoders that produce visual and textual features. A cross-modality spatio-temporal encoder which fuses information across spatial/temporal dimensions and visual/textual modalities. A language guided query selection module to initialize cross-modal queries. A cross-modality spatio-temporal decoder to decoder queries while fusing information from visual/textual features. Finally, two prediction heads predict the bounding boxes per frame and the temporal tube

aggregation modules, and initializes its spatial modules using pre-trained image-based representations. While the image and text feature extractors remain frozen, video-specific spatio-temporal adaptations are managed through learnable adapter blocks. This approach bridges the gap between image-based foundation models and the Spatio-Temporal Video Grounding (STVG) task by preserving the nuanced representations of the foundational model, thereby enhancing the model's ability to generalize to novel and unseen samples.

Weakly Supervised Methods

Since the fine-grained labeling process of localizing a tube (i.e., annotate a spatial region for each frame in videos) for STSGV is labor-intensive and complicated, some works considered the problem of weakly supervised spatio-temporal grounding where only aligned video-sentence pairs are provided without any fine-grained regional annotation.

DVSA [178] focuses on grounding nouns or pronouns to specific bounding boxes within video frames. It addresses the challenge of segment-level supervision that may not apply to individual frames due to off-screen or occluded objects. By using frame-wise loss weighting, the model estimates the likelihood of an object's presence in each frame. If the likelihood is high, it prioritizes matching loss; otherwise, it applies a penalty based on low confidence. This approach ensures more precise predictions for applicable frames while avoiding irrelevant proposals. Visual-semantic similarities and object appearance order in the video are used to estimate frame-wise object likelihood.

In contrast to DVSA, Chen's model [19] aims to ground natural sentences in videos, outputting a spatio-temporal tube. This is achieved through a multiple instance learning (MIL) approach. Instance proposals are first extracted from the video, and their features are matched with the sentence using an attentive interactor. The model is trained with a diversity loss and a ranking loss to improve semantic matching. During inference, the instance that best matches the sentence is selected as the grounding result.

Vis-Ctx [110] improves upon frame-level MIL by introducing contextual similarity, addressing limitations in previous models like DVSA. This similarity score is based on two principles: each query should focus on its most relevant frames, and non-object frames should have less impact. The contextual similarity accounts for the presence of objects across frames, making it more reliable. Additionally, the model introduces visual clustering to leverage temporal information, ensuring that regions with high query similarity also exhibit visual consistency across frames.

AWGU [22] formulates spatio-temporal grounding as two interconnected MIL problems. It grounds queries in region proposals within frames at the spatial level and in pertinent frames at the temporal level. This dual MIL framework improves accuracy by considering human activities associated with the query object, capturing both its physical state and spatial relations. To improve grounding results and narrow the search space, AWGU also incorporates activity-driven guidance into region proposals.

WSTOG [135] introduces a unified causal framework that focuses on learning de-convolutional object-dependent associations. By addressing object-independent associations, the framework enhances accuracy and robustness in video object grounding tasks. It establishes more precise associations between objects based on their specific characteristics and dependencies, leading to improved performance in grounding.

Winner [71] advances spatio-temporal grounding by incorporating context-dependent analysis through vision-language grammar induction. The model constructs a language decomposition tree in a bottom-up manner, fusing visual cues with the tree using a structural attention mechanism. This enables a hierarchical understanding of unstructured video data. WINNER's top-down feature backtracking ensures consistency, while a text reconstruction objective aligns visual and language components. The model narrows the modality gap between video and text, achieving better grounding of spatio-temporal tube proposals.

In summary, spatio-temporal sentence grounding significantly advances the challenge of localizing objects and instances in videos by considering both temporal boundaries and spatial localization across frames. Through the development of both fully supervised and weakly supervised methods, researchers have devised innovative strategies to enhance the performance of this task. Fully supervised approaches, initially focusing on multi-stage processes, have evolved towards end-to-end models that integrate spatial and temporal reasoning more efficiently. Meanwhile, weakly supervised methods tackle the labor-intensive annotation problem by leveraging less granular video-sentence pairs. As the field progresses, there is growing interest in leveraging foundation models and transformer architectures to improve generalization, especially for out-of-distribution and open-vocabulary samples. Future work in STSGV will likely continue refining these approaches, exploring new datasets, and enhancing model robustness to better handle the complexity and diversity of real-world videos.

2.3.3 Dense Events Grounding

Dense events grounding [4], also known as video paragraph grounding, aims to localize dense events within an untrimmed video based on a paragraph description. Compared to single-sentence inputs, paragraphs provide a contextualized description of multiple events in temporal order, as shown in Fig. 2.19. While it seems simple to apply single-event grounding models to each sentence, the temporal order of dense events within the video often aligns closely with their sequence in the descriptive paragraph. Ignoring these temporal clues can result in inaccurate localization of event boundaries. Therefore, models must be designed to exploit both temporal order and the semantic relations among dense events for more accurate event localization.

DepNet [4] is proposed to tackle this challenge. DepNet aims to adaptively aggregate temporal and semantic information of dense events into a compact

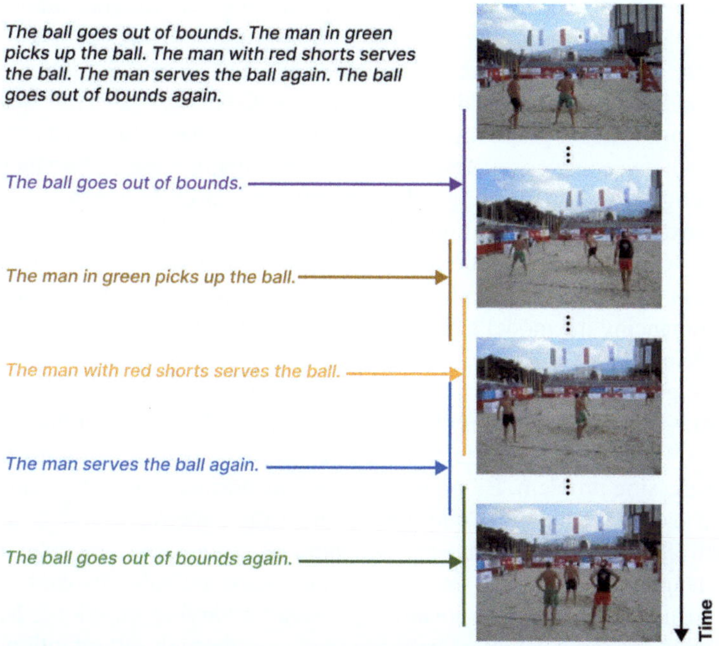

Fig. 2.19 Dense events grounding in video. Given a paragraph description, dense events grounding aims to localize described dense events in untrimmed video

set, which is then selectively propagated to each individual event. Specifically, DepNet generates visual-semantic moment proposals for each event described in the paragraph. Through its dense event aggregation module, it merges these moment proposals into a compact representation using second-order attention pooling. The dense events propagation module then selects features from the compact set and propagates them to each moment proposal via soft attention. This method enables the event proposals to incorporate temporal order and contextual relations from the paragraph, improving localization accuracy.

However, DepNet overlooks crucial contextual information in the paragraph and relies heavily on expensive temporal annotations. To address these limitations, SVPTR [58] is introduced. It incorporates contrastive learning and semi-supervised learning to reduce dependency on annotations while capturing hidden contextual information in paragraphs. The model extracts hierarchical text features and uses a sentence-based query mechanism in the decoder, allowing interaction among words and sentences to enhance contextual understanding. To prevent misalignment between proposal moments and sentences, contrastive learning is introduced to guide the cross-modal fusion at the video-paragraph level. Additionally, the teacher-student framework-based semi-supervised learning approach lessens the requirement for lengthy temporal annotations, increasing SVPTR's annotation efficiency.

Despite these advances, video paragraph grounding still presents challenges due to the complexity of the query input. Describing multiple events in a paragraph complicates word-level representation and can lead to misalignment between moments and sentences. To address these issues, GTLR [56] is proposed. GTLR introduces multi-modal graph modeling into the transformer to achieve fine-grained early fusion between video and text. Video clips and query words are connected with learnable edge weights, allowing selective information exchange. Using its long-range context modeling capabilities, the model then uses the transformer decoder to process sentence-level features and interactions between events. A language reconstructor further aids in rebuilding the paragraph query using multi-modal features, enhancing the model's ability to understand deep semantics and providing better explainability.

The fine-grained context for individual language queries may be lost as a result of DepNet's emphasis on the compact aggregation of temporal information. To resolve this, PRVG [111] introduces a new perspective, treating video grounding as a direct regression problem. By repurposing a Transformer-like architecture, PRVG employs a parallel decoding paradigm in which the paragraph directly attends to contextualized visual representations for moment regression. The cross-attention between language and visual representations enables PRVG to capture global context, ensuring flexibility in localizing events of varying durations. Moreover, PRVG models the mutual context between sentences through self-attention, helping to overcome semantic ambiguities. This simple yet effective design allows PRVG to perform grounding efficiently without requiring complex label assignments or post-processing.

Despite the strengths of these approaches, many methods still fail to fully leverage the hierarchical semantic correspondence between visual and textual modalities. To address this, HSCNet [123] is developed. HSCNet utilizes a multi-level encoder-decoder architecture to align visual and textual semantics across different levels. It progressively grounds queries from coarse to fine granularity, with lower-level queries conditioned on the knowledge from higher levels. This hierarchical approach enables the model to capture diverse temporal clues, which makes video paragraph grounding more accurate and fine-grained.

In contrast to previous models, SiamGTR [124] proposes a novel one-stage approach to weakly supervised video paragraph grounding, bypassing the need for generating proposals. SiamGTR employs two Siamese branches: an Augmentation Branch that processes pseudo videos with the complete paragraph as the language query for boundary supervision, and an Inference Branch that captures order-guided cross-modal correspondence from real videos. These branches share weights, allowing the model to jointly learn boundary prediction and feature association. This simple, efficient pipeline results in strong generalization, enabling SiamGTR to achieve superior performance even with weak supervision.

In summary, dense events grounding presents unique challenges due to the complexity of grounding multiple temporally ordered events within an untrimmed video. Models like DepNet have introduced innovative mechanisms for aggregating temporal and semantic information, but limitations such as reliance on expensive annotations and loss of fine-grained context persist. Subsequent models, including SVPTR, GTLR, PRVG, HSCNet, and SiamGTR, have sought to overcome these challenges through techniques like contrastive learning, graph-based multi-modal fusion, hierarchical semantic alignment, and weakly supervised learning. These advancements represent significant steps forward, yet the task remains difficult, as achieving precise alignment between dense events and descriptive paragraphs continues to push the boundaries of video grounding research.

2.3.4 Video-Language Pre-training for Temporal Grounding

Video-language pre-training [86, 102] has been shown to enhance many downstream text-based video understanding tasks, such as video captioning, video question answering, and video retrieval. However, these models primarily focus on understanding entire videos and are less effective for tasks like *Video Grounding*, which require a more granular understanding of specific video segments. To address this limitation, large-scale video-text datasets with precise temporal boundary annotations, along with temporally-aware training losses, are necessary.

Existing large-scale video-text datasets, such as WebVID [1], often lack temporal annotations, while others like HowTo100M [92] suffer from noisy temporal data. This gap hinders the development of models that can effectively perform fine-grained temporal alignment, which is critical for accurate video understanding. To tackle this challenge, several pre-training methods have been proposed to improve models' ability to localize temporal boundaries within videos.

BSP [145] introduces an innovative approach to pre-training for temporal localization. Unlike traditional methods that rely on expensive manual annotations, BSP synthesizes temporal boundaries using action classification datasets. It accomplishes this by outlining different approaches for drawing these boundaries, enabling BSP to operate through boundary-type classification in a self-supervised fashion. This approach helps models learn video representations that can be easily transferred to downstream temporal localization tasks including TAL, *Video Grounding*, and step localization.

Building on this idea, TemPVL [91] is developed to enable fine-grained temporal and semantic alignment. TemPVL focuses on two key tasks: moment retrieval, which predicts the start and end boundaries in videos based on textual descriptions, and text localization, which aligns text segments with video features. The method manually concatenates frame features from multiple videos into a long sequence, which interacts with a corresponding text sequence. This method enables the model to differentiate between various instances within each modality by connecting fine-grained frame representations with word representations. This enhances the model's ability to recognize and differentiate temporal boundaries within videos.

While BSP synthesizes boundaries using action recognition datasets, LocVTP [9] extends this concept by introducing a pre-training framework specifically for the video-language domain. LocVTP moves beyond the label-dependent pre-training

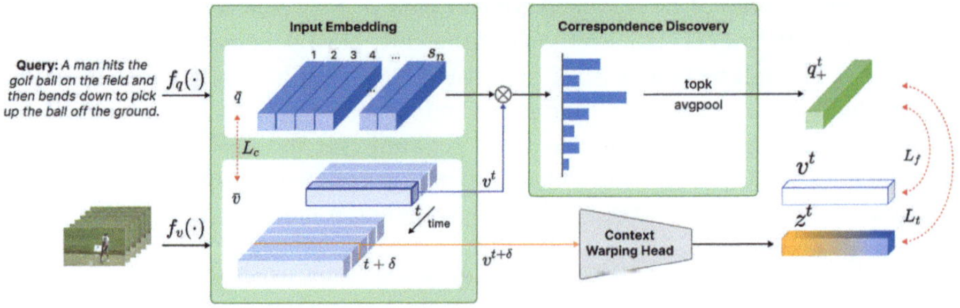

Fig. 2.20 The architecture of LocVTP. It consists of: (1) Coarse-grained contrastive loss matches the global video and sentence representations. (2) The clip-word correspondence is firstly built by similarity computing and then fine-grained contrastive loss conducts detailed cross-modal alignment. (3) A context warping head is employed to warp the contextual feature and a temporal aware contrastive loss is applied based on the warped feature

of BSP by directly utilizing narrated videos. As shown in Fig. 2.20, it employs a fine-grained contrastive alignment strategy through a clip-word correspondence discovery mechanism, which strengthens the model's temporal reasoning abilities. Additionally, LocVTP incorporates a context projection head and a temporal-aware contrastive loss, improving the model's understanding of contextual relationships and providing a more nuanced approach to temporal localization.

Zeng et al. [159] introduce the graph pre-training upon their multi-modal relational graph to enhance the visual features with explicit relations. They design two node-level and graph-level self-supervised pre-training tasks (i.e., attribute masking and cross-modal context prediction).

Recently, Wang et al. [136] introduce InternVideo2, a video annotation framework based on a multi-modal large language model, which generates fine-grained textual annotations for videos, creating a large-scale, high-quality video-text dataset. InternVideo2 follows a three-stage training process, with each stage producing a model used for different downstream tasks. Specifically, the first stage focuses on video representation reconstruction, training the video encoder for spatiotemporal understanding. The second stage employs video-audio-text multi-modal contrastive learning to enhance cross-modal matching capabilities. In the third stage, the video encoder is aligned with a large language model, resulting in a powerful video-language model. The video encoder trained in the second stage is used to extract video features for the CG-DETR model. Compared to the previous CLIP+SlowFast features, it shows a performance improvement of over 10%, making it one of the strongest models for the *Video Grounding* task.

These models primarily focus on video feature extraction, where the extracted features replace those from models like C3D or CLIP, while keeping the architecture of *Video Grounding* models unchanged. In contrast, Lin et al. [73] propose UniVTG, which introduces a new model architecture that unifies various video-text grounding tasks, including video grounding, video highlight detection, and video summarization. UniVTG employs a flexible model and label scaling approach to handle different task annotations. The model leverages point annotations from the Ego4D [41] dataset and segment annotations from the VideoCC [98] dataset, using image-text similarity scores computed by the CLIP model as guidance. It undergoes large-scale pre-training and is fine-tuned on various video grounding downstream tasks, achieving impressive results.

In conclusion, video-language pre-training has demonstrated remarkable potential in enhancing temporal grounding tasks by addressing the need for precise temporal boundary annotations and fine-grained alignment. Approaches like BSP, TemPVL, LocVTP, and graph-based methods have made significant strides in improving models' ability to localize events within videos, with techniques ranging from self-supervised boundary synthesis to fine-grained contrastive alignment. The introduction of large-scale datasets like InternVideo2 and unified architectures such as UniVTG further expand the horizon for video-language pre-training by enabling the development of more robust, versatile models. Despite these developments, there is still much room for new developments in video grounding research because temporal reasoning and localization in videos are still difficult.

2.3.5 Long-Form Video Grounding

Long-form video grounding [114] is an advanced task in video grounding, focusing on aligning natural language queries with relevant moments in extended video sequences. Unlike traditional video grounding, which typically deals with short videos under 3 minutes, long-form video grounding handles multi-event videos of much longer durations. This introduces challenges in temporal reasoning, context comprehension, and managing interleaved events. Common datasets like Ego4D-NLQ [41] (average 8 minutes) and MAD [114] (over 1 hour) serve as benchmarks, testing models' abilities to understand complex, lengthy video content. This task is significant as it pushes video understanding closer to real-world scenarios, where videos are often lengthy and involve numerous interconnected events.

Early approaches to long-form video grounding extended short video grounding techniques by either downsampling videos or applying a sliding window strategy. However, these methods caused temporal information loss and increased computational costs due to the sheer number of candidate windows in long videos. Sparse sampling, for instance, reduced visible frames, limiting fine-grained content alignment. To address this, CONE [48] introduces a Coarse-to-Fine alignment strategy. It significantly reduces the number of candidate windows by initially segmenting videos into potential segments and subsequently selecting relevant windows based on specific queries. A coarse-to-fine mechanism progressively refines the alignment of multi-modal inputs across different granularities. This approach enhances computational efficiency and improves multi-modal alignment, particularly through the integration of contrastive learning techniques and vision-text models such as CLIP, thereby achieving superior performance.

While CONE improves efficiency, its sparse sampling approach may overlook the full potential of long-form videos. SOONet [103], an anchor-based end-to-end framework, offers a solution by scanning the entire long-form video only once. It divides the video into non-overlapping anchor sequences and implements a three-step procedure for accurate predictions: multi-scale context modeling, content-based anchor re-ranking, and boundary regression. This method reduces computational complexity while enhancing the cross-modal alignment between video and query. By simultaneously grounding several queries in a single video and optimizing the learning process with a dual-form rank loss that improves both intra-anchor and inter-anchor knowledge, SOONet further increases efficiency.

Barrios et al. [5] highlight a critical limitation: certain video segments are harder to describe with natural language, which leads to noisy predictions in long videos. To address this, they propose a two-stage guided grounding framework. The first stage, the Guidance Model, identifies the "describable" windows by leveraging multi-modal cues such as audio and visual signals. The base grounding model is used in the second stage, processing brief windows to correspond with particular queries. This approach can operate in two modes: query agnostic, for real-time applications, and query dependent, for higher precision at the cost of computational resources. The flexible nature of this framework allows it to be adapted to various grounding methods, enhancing its utility across different applications.

A common challenge in video activity localization is the uncertainty of action boundaries, which are subjective and ambiguous. Drawing inspiration from object detection, DenoiseLoc [146] introduces a boundary-denoising training paradigm. Its encoder-decoder architecture iteratively refines temporal predictions, progressively improving boundary accuracy. By introducing noisy ground-truth spans during training, DenoiseLoc forces the model to predict more reliable action boundaries. Despite utilizing significantly fewer proposals, this denoising technique performs on par with more computationally demanding models and generalizes across a variety of video domains.

Existing long-form video grounding models often separate clip retrieval from grounding, which hampers their fine-grained event understanding. RGNet [43] addresses this issue by unifying both tasks within a single network, enabling end-to-end training. By optimizing the retrieval module with moment annotations, RGNet enhances both clip retrieval and moment grounding. It achieves this through a novel RG-Encoder, which integrates sparse attention mechanisms to focus on key events within the video. The use of intra-clip attention and negative clip mining further refines the model's ability to distinguish relevant clips from irrelevant ones, making it more effective for long-form video tasks.

Many existing models struggle with the computational costs of processing long videos. SnAG [96] tackles this problem by adopting a late-fusion design, which amortizes the processing cost across multiple text queries. This design, combined with a video-centric sampling scheme for training, allows SnAG to scale effectively without sacrificing accuracy. Despite its simplicity, SnAG delivers superior performance on long-form video grounding benchmarks, demonstrating the value of its efficient and scalable approach.

In summary, long-form video grounding represents a significant evolution in the field of video grounding, addressing the complexities inherent in longer, multi-event video sequences. As models such as CONE, SOONet, and RGNet advance the efficiency and precision of temporal alignment, they offer promising solutions to the challenges posed by extensive video content. Techniques like boundary denoising, coarse-to-fine alignment, and late-fusion design demonstrate innovative approaches to tackling the issues of computational cost, temporal ambiguity, and multi-query processing. However, the continuous development of more robust models is crucial to fully capture the intricate dynamics of long-form videos, pushing video understanding closer to real-world applications.

2.3.6 Video Corpus Grounding

Video Corpus Grounding (VCG) [32], also referred to as video corpus moment retrieval, is the task of identifying specific video segments from a large collection that match a given query. Usually, this challenge is broken down into two subtasks: single video grounding (SVG), which entails identifying the precise time segment within a chosen video, and video retrieval (VR), where the model searches through a corpus to find videos that might contain the relevant segment (also known as partially relevant video retrieval [30]).

Initially, VCG is addressed by adapting existing SVG datasets such as ActivityNet-Captions, Didemo, and Charades-STA to the video corpus grounding task. CAL [32] addresses VCG by representing a video moment as a sequence of short clips and aligning these clips with the natural language query. A key innovation in CAL is the clip-alignment cost, which measures the similarity between language and video clip features using squared-Euclidean distance. This fine-grained alignment allows for efficient video indexing and retrieval, making it particularly effective for temporal alignment at a clip level. The model operates in two stages: it first retrieves candidate clips using an approximate nearest-neighbor search and then refines the results by applying the full alignment cost to nearby moments. While effective, CAL relies on predefined clip proposals, which can limit precision, particularly for long videos.

HAMMER [163] focuses on balancing the demands of retrieving long videos and localizing short segments by encoding videos at multiple levels of granularity. The model leverages coarse-grained representations for long video spans and fine-grained representations for frame-level precision. Using a hierarchical cross-modal attention mechanism, HAMMER aligns text and visual features at frame, clip, and video levels. While this structure allows for more precise video segment localization: the cross-modal interaction with each video in the corpus increases computational complexity.

Lei et al. [68] collect the TVR dataset for VCMR, which consists of 21,793 videos and 109,000 queries from six TV shows, in the genres of sitcom, medical drama, and crime. Building on CAL's limitations, XML is developed to address inefficiencies related to predefined proposals and high computational cost. XML employs a late fusion approach where videos and queries are encoded independently, significantly reducing computation at test time since videos can be pre-encoded. By integrating these representations with matrix multiplication, XML generates similarity scores over the temporal dimension of videos. The model also introduces a convolutional start-end detector to refine moment predictions, resulting in better performance than previous proposal-based methods. XML's computational efficiency and end-to-end trainability render it a robust improvement over earlier models.

ReLoCLNet [170] employs unimodal encoding, where video and text are encoded separately. The innovation in ReLoCLNet lies in introducing contrastive learning to enhance cross-modal interactions. By training the model with VideoCL and FrameCL objectives, it strengthens the relationship between semantically related video segments and queries. This approach allows for the efficient pre-encoding of video features while achieving fine-grained localization through frame-level contrastive learning.

Despite improvements, existing VCG models still suffer from retrieval bias, where certain objects in the query (e.g., "coffee") are consistently associated with specific scenes (e.g., "drinking"). SQuiDNet [152] addresses this issue by disentangling "good" and "bad" biases in retrieval predictions. It introduces two retrieval models: Naive Moment Retrieval and Biased Moment Retrieval, which work together to mitigate retrieval bias. SQuiDNet learns to selectively debias by understanding the meaning of the query, improving accuracy in identifying relevant video segments while reducing the influence of spurious correlations.

The frame-based retrieval approaches in previous models often fail to capture the semantic structure inherent in video events. In order to solve this, EventFormer [49] focuses on event-level representations rather than frame-level ones. Drawing from cognitive science, which suggests that humans process visual information in events, EventFormer segments videos into coherent events. By explicitly modeling event-level interactions, the model improves retrieval accuracy and efficiency. Additionally, EventFormer introduces novel training strategies, including contrastive learning and hierarchical event encoding, to enhance both video retrieval and single video moment localization.

HERO [70] is a pre-trained video-text model fine-tuned for the video corpus grounding tasks. The model takes video frames and their associated subtitles as input and utilizes a hierarchical structure for multi-modal encoding. First, a Cross-modal Transformer fuses the subtitle and corresponding video frames at the local level, followed by a Temporal Transformer that captures the global temporal context by considering all surrounding frames. This design improves performance over simpler BERT-like models, as it better handles the sequential nature of video data. HERO employs four pre-training tasks: Masked Language Modeling (MLM), Masked Frame Modeling (MFM), Video-Subtitle Matching (VSM), and Frame Order Modeling (FOM). Particularly novel is the introduction of VSM and FOM, which guarantee both local and global temporal alignment between text and video.

Extensive experiments confirm that these tasks significantly enhance video-text alignment during pre-training.

Following HERO, some models adopted a two-stage approach, using HERO for video retrieval and training a separate grounding model only for the top-ranked videos as following:

CONQUER [47] addresses limitations in query-dependent fusion (QDF) and query-aware feature learning (QAL), which have been underexplored in earlier models. The task of video corpus grounding demands not just global video retrieval but also precise moment localization, making the integration of QDF and QAL critical. In CONQUER, QDF derives query-specific fusion weights to extract relevant video features, while QAL further refines these features by coupling them with the query for retrieval. This two-step process creates a user-centered representation that progressively refines video and query features, allowing for a more detailed understanding of diverse queries. This advancement highlights the need for better interaction between query and video contexts, which earlier models like HERO did not fully address.

While earlier models have made progress in fusing visual and textual information, they often fail to account for different levels of relevance between the two modalities. CKCN [26] introduces a novel approach to overcome these shortcomings by dynamically utilizing the most relevant modality based on the query. The model identifies whether a query is related to subtitles, video frames, or both, and adjusts its feature fusion accordingly. Moreover, CKCN tackles the issue of redundant information from co-attention mechanisms, which can introduce noise and weaken the alignment between video and query. Using a dual calibration transformer, the Cross-Modality Knowledge Calibration Network reduces superfluous information and directs multi-modal fusion according to the semantic content of the query. This approach significantly improves cross-modality interactions and moment localization accuracy.

Despite advancements in models like CKCN, video feature extraction often remains coarse, failing to capture fine-grained details necessary for accurate video corpus grounding.

DMFAT [173] addresses this by incorporating multi-granularity feature extraction, using deformable attention in the encoder to capture both detailed local features and abstract global semantics. This multi-level approach ensures more comprehensive information extraction. Additionally, DMFAT enhances the standard Transformer with deformation attention, enabling the model to capture structural nuances in video sequences. A novel adversarial training strategy combining contrastive learning with adversarial attacks is introduced to further align video and query features. By incorporating both video-level and frame-level contrastive learning, DMFAT achieves improved accuracy in locating the correct video segments, refining the overall process of video grounding.

In summary, video corpus grounding has evolved significantly through the development of various models, each addressing different challenges such as retrieval bias, computational complexity, and the fine-grained alignment of video segments with textual queries. Early approaches like CAL and XML focus on efficient clip-level matching and reducing the reliance on predefined proposals, while more recent models such as SQuiDNet and EventFormer have shifted attention to debiased and event-level representations, respectively. Models like HERO introduced innovative pre-training tasks and advanced query-dependent fusion techniques, improving both video retrieval and moment localization. Meanwhile, CKCN and DMFAT have pushed the boundaries of cross-modality interactions and feature granularity, enhancing both precision and semantic understanding. Even with these developments, it is still difficult to strike a better balance between fine-grained moment localization and computational efficiency. Future research will probably examine increasingly complex techniques for incorporating cross-modal features and improving retrieval accuracy as the field develops.

2.3.7 Identity-Text Video Corpus Grounding

In this section, we introduce the Identity-Text Video Corpus Grounding (ITVCG) task [53]. In all previously discussed *Video Grounding* tasks, the query is purely text-based. However, in situations where visual identity is essential for retrieval, textual descriptions might not be enough to fully capture the complexity of user queries.

Imagine a scenario where a user wants to retrieve a video moment featuring a specific person performing an action. Conventional video corpus grounding (VCG) models may have trouble differentiating between people in a scene because they only use textual descriptions. For instance, given the query "Castle and Beckett are slow dancing together", traditional models can identify moments of slow dancing but may fail to ensure that the retrieved moment contains the correct individuals. To address this limitation, ITVCG incorporates both textual descriptions and visual identity images into the query. As illustrated in Fig. 2.21, this additional modality significantly improves retrieval accuracy and enables more flexible search capabilities.

To support ITVCG research, the TVR-IT dataset is constructed as an extension of the TVR dataset, incorporating 463 identity images from six TV shows. At least one identity image can be found in 68,840 (94.5%) of the 72,840 queries in this dataset. Additionally, identity-seen and identity-unseen dataset splits are introduced to evaluate the generalization ability of models in handling unseen identities.

To tackle the ITVCG task, Video-Locator [53] is proposed as the first model specifically designed for this challenge. As shown in Fig. 2.22, Video-Locator consists of two core components.

- Video-Identity-Text Alignment—This module processes videos, identity images, and textual queries separately and aligns their representations using contrastive learning. By doing so, the model ensures that video moments containing the correct identities are accurately retrieved.
- Multi-Modal Fine-Grained Fusion—After retrieving a relevant video, this module refines the search by analyzing the retrieved video in conjunction with the given query. It generates precise timestamps, identifying the start and end of the relevant moment.

Fig. 2.21 Traditional video corpus grounding fails to identify identities within videos. Identity-Text Video Corpus Grounding task incorporates visual identities into the query, enabling more precise and flexible video grounding

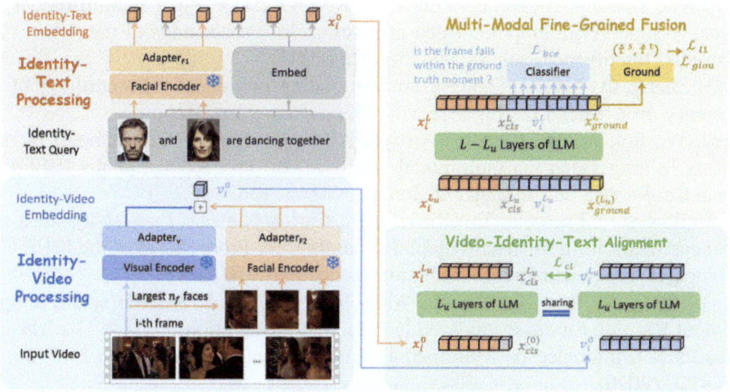

Fig. 2.22 The framework of the Video-Locator model. It introduces separate facial adapters and a visual Adapter to enable the LLM to comprehend facial and visual features. Subsequently, Identity-Text and Identity-Video features are aligned through the Video-Identity-Text Alignment module. The Multi-Modal Fine-Grained Fusion module integrates representations of matched videos and queries to generate timestamps

Despite the promising results achieved by Video-Locator, the ITVCG task presents unique challenges, particularly in terms of identity generalization. Existing models often struggle to recognize identities beyond those seen in training, which limits their applicability in real-world scenarios. Addressing this issue requires further research, including large-scale identity-pre-training strategies and improved representation learning techniques.

By combining textual and visual information, ITVCG offers users more accurate and adaptable retrieval options, making it a significant advancement in video grounding. Moving forward, multi-modal queries should be further explored to enhance video grounding. While ITVCG introduces identity-based queries, other modalities, such as audio and motion cues, could provide richer contextual information for retrieval. Developing more flexible and adaptive grounding models that can seamlessly integrate various modalities will be essential for advancing video understanding and retrieval systems.

References

1. Bain, M., Nagrani, A., Varol, G., Zisserman, A.: Frozen in time: a joint video and image encoder for end-to-end retrieval. In: Proceedings of the IEEE/CVF International Conference on Computer Vision, pp. 1728–1738 (2021)
2. Bao, P., Shao, Z., Yang, W., Ng, B.P., Er, M.H., Kot, A.C.: Omnipotent distillation with llms for weakly-supervised natural language video localization: when divergence meets consistency. In: Proceedings of the AAAI Conference on Artificial Intelligence, vol. 38, pp. 747–755 (2024)
3. Bao, P., Xia, Y., Yang, W., Ng, B.P., Er, M.H., Kot, A.C.: Local-global multi-modal distillation for weakly-supervised temporal video grounding. In: Proceedings of the AAAI Conference on Artificial Intelligence, vol. 38, pp. 738–746 (2024)
4. Bao, P., Zheng, Q., Mu, Y.: Dense events grounding in video. In: Proceedings of the AAAI Conference on Artificial Intelligence, vol. 35, pp. 920–928 (2021)
5. Barrios, W., Soldan, M., Ceballos-Arroyo, A.M., Heilbron, F.C., Ghanem, B.: Localizing moments in long video via multimodal guidance. In: Proceedings of the IEEE/CVF International Conference on Computer Vision, pp. 13667–13678 (2023)
6. Caba Heilbron, F., Escorcia, V., Ghanem, B., Carlos Niebles, J.: Activitynet: a large-scale video benchmark for human activity understanding. In: Proceedings of the IEEE Conference on Computer Vision and Pattern Recognition, pp. 961–970 (2015)
7. Cao, D., Zeng, Y., Liu, M., He, X., Wang, M., Qin, Z.: STRONG: spatio-temporal reinforcement learning for cross-modal video moment localization. In: MM '20: The 28th ACM International Conference on Multimedia, Virtual Event, Seattle, October 12–16, 2020, pp. 4162–4170 (2020)
8. Cao, D., Zeng, Y., Wei, X., Nie, L., Hong, R., Qin, Z.: Adversarial video moment retrieval by jointly modeling ranking and localization. In: MM '20: The 28th ACM International Conference on Multimedia, Virtual Event, Seattle, October 12–16, 2020, pp. 898–906 (2020)
9. Cao, M., Yang, T., Weng, J., Zhang, C., Wang, J., Zou, Y.: Locvtp: video-text pre-training for temporal localization. In: European Conference on Computer Vision, pp. 38–56. Springer (2022)
10. Cao, M., Wei, F., Xu, C., Geng, X., Chen, L., Zhang, C., Zou, Y., Shen, T., Jiang, D.: Iterative proposal refinement for weakly-supervised video grounding. In: Proceedings of the IEEE/CVF Conference on Computer Vision and Pattern Recognition, pp. 6524–6534 (2023)
11. Carion, N., Massa, F., Synnaeve, G., Usunier, N., Kirillov, A., Zagoruyko, S.: End-to-end object detection with transformers. In: European Conference on Computer Vision, pp. 213–229. Springer (2020)
12. Chen, C., Gu, X.: Semantic modulation based residual network for temporal language queries grounding in video. In: International Symposium on Neural Networks, pp. 119–129. Springer (2020)

13. Chen, S., Jiang, Y.G.: Semantic proposal for activity localization in videos via sentence query. In: Proceedings of the AAAI Conference on Artificial Intelligence, vol. 33, pp. 8199–8206 (2019)
14. Chen, S., Jiang, Y.G.: Hierarchical visual-textual graph for temporal activity localization via language. In: European Conference on Computer Vision, pp. 601–618. Springer (2020)
15. Chen, S., Jiang, Y.G.: Towards bridging event captioner and sentence localizer for weakly supervised dense event captioning. In: Proceedings of the IEEE/CVF Conference on Computer Vision and Pattern Recognition, pp. 8425–8435 (2021)
16. Chen, D., Fisch, A., Weston, J., Bordes, A.: Reading Wikipedia to answer open-domain questions. In: Proceedings of the 55th Annual Meeting of the Association for Computational Linguistics (Volume 1: Long Papers), pp. 1870–1879 (2017)
17. Chen, J., Chen, X., Ma, L., Jie, Z., Chua, T.S.: Temporally grounding natural sentence in video. In: Proceedings of the 2018 Conference on Empirical Methods in Natural Language Processing, pp. 162–171 (2018)
18. Chen, J., Ma, L., Chen, X., Jie, Z., Luo, J.: Localizing natural language in videos. In: Proceedings of the AAAI Conference on Artificial Intelligence, vol. 33, pp. 8175–8182 (2019)
19. Chen, Z., Ma, L., Luo, W., Wong, K.Y.K.: Weakly-supervised spatio-temporally grounding natural sentence in video. In: Proceedings of the 57th Annual Meeting of the Association for Computational Linguistics, pp. 1884–1894 (2019)
20. Chen, L., Lu, C., Tang, S., Xiao, J., Zhang, D., Tan, C., Li, X.: Rethinking the bottom-up framework for query-based video localization. In: Proceedings of the AAAI Conference on Artificial Intelligence, vol. 34, pp. 10551–10558 (2020)
21. Chen, S., Jiang, W., Liu, W., Jiang, Y.G.: Learning modality interaction for temporal sentence localization and event captioning in videos. In: Computer Vision—ECCV 2020: 16th European Conference, Glasgow, August 23–28, 2020, Proceedings, Part IV 16, pp. 333–351. Springer (2020)
22. Chen, J., Bao, W., Kong, Y.: Activity-driven weakly-supervised spatio-temporal grounding from untrimmed videos. In: Proceedings of the 28th ACM International Conference on Multimedia, pp. 3789–3797 (2020)
23. Chen, Z., Ma, L., Luo, W., Tang, P., Wong, K.Y.K.: Look closer to ground better: weakly-supervised temporal grounding of sentence in video. ArXiv preprint abs/2001.09308 (2020)
24. Chen, J., Luo, W., Zhang, W., Ma, L.: Explore inter-contrast between videos via composition for weakly supervised temporal sentence grounding. In: Proceedings of the AAAI Conference on Artificial Intelligence, vol. 36, pp. 267–275 (2022)
25. Chen, H., Wang, X., Chen, H., Song, Z., Jia, J., Zhu, W.: Grounding-prompter: prompting llm with multimodal information for temporal sentence grounding in long videos. arXiv preprint arXiv:2312.17117 (2023)
26. Chen, T., Wang, W., Jiang, Z., Li, R., Wang, B.: Cross-modality knowledge calibration network for video corpus moment retrieval. IEEE Trans. Multimedia. **26**, 3799–3813 (2023)
27. Cui, R., Qian, T., Peng, P., Daskalaki, E., Chen, J., Guo, X., Sun, H., Jiang, Y.G.: Video moment retrieval from text queries via single frame annotation. In: Proceedings of the 45th International ACM SIGIR Conference on Research and Development in Information Retrieval, pp. 1033–1043 (2022)
28. Ding, X., Wang, N., Zhang, S., Cheng, D., Li, X., Huang, Z., Tang, M., Gao, X.: Support-set based cross-supervision for video grounding. In: Proceedings of the IEEE/CVF International Conference on Computer Vision, pp. 11573–11582 (2021)
29. Diwan, A., Peng, P., Mooney, R.: Zero-shot video moment retrieval with off-the-shelf models. In: Transfer Learning for Natural Language Processing Workshop, pp. 10–21. PMLR (2023)
30. Dong, J., Chen, X., Zhang, M., Yang, X., Chen, S., Li, X., Wang, X.: Partially relevant video retrieval. In: Proceedings of the 30th ACM International Conference on Multimedia, pp. 246–257 (2022)
31. Duan, X., Huang, W., Gan, C., Wang, J., Zhu, W., Huang, J.: Weakly supervised dense event captioning in videos. In: Advances in Neural Information Processing Systems 31: Annual Conference on Neural Information Processing Systems 2018, NeurIPS 2018, December 3–8, 2018, Montréal, pp. 3063–3073 (2018)

32. Escorcia, V., Soldan, M., Sivic, J., Ghanem, B., Russell, B.: Temporal localization of moments in video collections with natural language. arXiv preprint arXiv:1907.12763 (2019)
33. Feng, W., Wang, X., Chen, H., Zhang, Z., Song, Z., Zhou, Y., Zhu, W.: Llm4vg: large language models evaluation for video grounding. arXiv preprint arXiv:2312.14206 (2023)
34. Gao, J., Xu, C.: Fast video moment retrieval. In: Proceedings of the IEEE/CVF International Conference on Computer Vision, pp. 1523–1532 (2021)
35. Gao, J., Xu, C.: Learning video moment retrieval without a single annotated video. IEEE Trans. Circuits Syst. Video Technol. **32**(3), 1646–1657 (2021)
36. Gao, J., Sun, C., Yang, Z., Nevatia, R.: TALL: temporal activity localization via language query. In: IEEE International Conference on Computer Vision, ICCV 2017, Venice, October 22–29, 2017, pp. 5277–5285 (2017)
37. Gao, M., Davis, L., Socher, R., Xiong, C.: WSLLN: weakly supervised natural language localization networks. In: Proceedings of the 2019 Conference on Empirical Methods in Natural Language Processing and the 9th International Joint Conference on Natural Language Processing (EMNLP-IJCNLP), pp. 1481–1487 (2019)
38. Ge, R., Gao, J., Chen, K., Nevatia, R.: Mac: mining activity concepts for language-based temporal localization. In: 2019 IEEE Winter Conference on Applications of Computer Vision (WACV), pp. 245–253. IEEE (2019)
39. Ghosh, S., Agarwal, A., Parekh, Z., Hauptmann, A.: ExCL: extractive clip localization using natural language descriptions. In: Proceedings of the 2019 Conference of the North American Chapter of the Association for Computational Linguistics: Human Language Technologies, Volume 1 (Long and Short Papers), pp. 1984–1990 (2019)
40. Girshick, R.B., Donahue, J., Darrell, T., Malik, J.: Rich feature hierarchies for accurate object detection and semantic segmentation. In: 2014 IEEE Conference on Computer Vision and Pattern Recognition, CVPR 2014, Columbus, June 23–28, 2014, pp. 580–587 (2014)
41. Grauman, K., Westbury, A., Byrne, E., Chavis, Z., Furnari, A., Girdhar, R., Hamburger, J., Jiang, H., Liu, M., Liu, X., et al.: Ego4d: around the world in 3,000 hours of egocentric video. In: Proceedings of the IEEE/CVF Conference on Computer Vision and Pattern Recognition, pp. 18995–19012 (2022)
42. Hahn, M., Kadav, A., Rehg, J.M., Graf, H.P.: Tripping through time: efficient localization of activities in videos. In: 31st British Machine Vision Conference 2020, BMVC 2020, Virtual Event, September 7–10, 2020 (2020)
43. Hannan, T., Islam, M.M., Seidl, T., Bertasius, G.: Rgnet: a unified retrieval and grounding network for long videos. arXiv preprint arXiv:2312.06729 (2023)
44. He, D., Zhao, X., Huang, J., Li, F., Liu, X., Wen, S.: Read, watch, and move: reinforcement learning for temporally grounding natural language descriptions in videos. In: Proceedings of the AAAI Conference on Artificial Intelligence, vol. 33, pp. 8393–8400 (2019)
45. Hendricks, L.A., Wang, O., Shechtman, E., Sivic, J., Darrell, T., Russell, B.C.: Localizing moments in video with natural language. In: IEEE International Conference on Computer Vision, ICCV 2017, Venice, October 22–29, 2017, pp. 5804–5813 (2017)
46. Hendricks, L.A., Wang, O., Shechtman, E., Sivic, J., Darrell, T., Russell, B.: Localizing moments in video with temporal language. arXiv preprint arXiv:1809.01337 (2018)
47. Hou, Z., Ngo, C.W., Chan, W.K.: Conquer: contextual query-aware ranking for video corpus moment retrieval. In: Proceedings of the 29th ACM International Conference on Multimedia, pp. 3900–3908 (2021)
48. Hou, Z., Zhong, W., Ji, L., Gao, D., Yan, K., Chan, W.K., Ngo, C.W., Shou, Z., Duan, N.: Cone: an efficient coarse-to-fine alignment framework for long video temporal grounding. arXiv preprint arXiv:2209.10918 (2022)
49. Hou, D., Pang, L., Shen, H., Cheng, X.: Event-aware video corpus moment retrieval. arXiv preprint arXiv:2402.13566 (2024)
50. Hu, R., Xu, H., Rohrbach, M., Feng, J., Saenko, K., Darrell, T.: Natural language object retrieval. In: 2016 IEEE Conference on Computer Vision and Pattern Recognition, CVPR 2016, Las Vegas, June 27–30, 2016, pp. 4555–4564 (2016)

51. Huang, J., Liu, Y., Gong, S., Jin, H.: Cross-sentence temporal and semantic relations in video activity localisation. In: Proceedings of the IEEE/CVF International Conference on Computer Vision, pp. 7199–7208 (2021)
52. Huang, Y., Yang, L., Sato, Y.: Weakly supervised temporal sentence grounding with uncertainty-guided self-training. In: Proceedings of the IEEE/CVF Conference on Computer Vision and Pattern Recognition, pp. 18908–18918 (2023)
53. Huang, B., Wang, X., Chen, H., Chen, H., Wu, Y., Zhu, W.: Identity-text video corpus grounding (2025)
54. Jang, J., Park, J., Kim, J., Kwon, H., Sohn, K.: Knowing where to focus: event-aware transformer for video grounding. In: Proceedings of the IEEE/CVF International Conference on Computer Vision, pp. 13846–13856 (2023)
55. Jiang, B., Huang, X., Yang, C., Yuan, J.: Cross-modal video moment retrieval with spatial and language-temporal attention. In: Proceedings of the 2019 on International Conference on Multimedia Retrieval, pp. 217–225 (2019)
56. Jiang, X., Xu, X., Zhang, J., Shen, F., Cao, Z., Cai, X.: Gtlr: graph-based transformer with language reconstruction for video paragraph grounding. In: 2022 IEEE International Conference on Multimedia and Expo (ICME), pp. 1–6. IEEE (2022)
57. Jiang, X., Xu, X., Zhang, J., Shen, F., Cao, Z., Shen, H.T.: Semi-supervised video paragraph grounding with contrastive encoder. In: Proceedings of the IEEE/CVF Conference on Computer Vision and Pattern Recognition, pp. 2466–2475 (2022)
58. Jiang, X., Xu, X., Zhang, J., Shen, F., Cao, Z., Shen, H.T.: Semi-supervised video paragraph grounding with contrastive encoder. In: Proceedings of the IEEE/CVF Conference on Computer Vision and Pattern Recognition, pp. 2466–2475 (2022)
59. Jiang, X., Zhou, Z., Xu, X., Yang, Y., Wang, G., Shen, H.T.: Faster video moment retrieval with point-level supervision. arXiv preprint arXiv:2305.14017 (2023)
60. Jin, Y., Yuan, Z., Mu, Y., et al.: Embracing consistency: a one-stage approach for spatio-temporal video grounding. Adv. Neural Inf. Process. Syst. **35**, 29192–29204 (2022)
61. Jung, M., Jang, Y., Choi, S., Kim, J., Kim, J.H., Zhang, B.T.: Overcoming weak visual-textual alignment for video moment retrieval. arXiv preprint arXiv:2306.02728 (2023)
62. Kim, D., Park, J., Lee, J., Park, S., Sohn, K.: Language-free training for zero-shot video grounding. In: Proceedings of the IEEE/CVF Winter Conference on Applications of Computer Vision, pp. 2539–2548 (2023)
63. Kim, S., Cho, J., Yu, J., Yoo, Y., Choi, J.Y.: Gaussian mixture proposals with pull-push learning scheme to capture diverse events for weakly supervised temporal video grounding. In: Proceedings of the AAAI Conference on Artificial Intelligence, vol. 38, pp. 2795–2803 (2024)
64. Kipf, T.N., Welling, M.: Semi-supervised classification with graph convolutional networks. In: 5th International Conference on Learning Representations, ICLR 2017, Toulon, April 24–26, 2017, Conference Track Proceedings (2017)
65. Kong, S., Li, L., Zhang, B., Wang, W., Jiang, B., Yan, C., Xu, C.: Dynamic contrastive learning with pseudo-samples intervention for weakly supervised joint video mr and hd. In: Proceedings of the 31st ACM International Conference on Multimedia, pp. 538–546 (2023)
66. Krishna, R., Hata, K., Ren, F., Fei-Fei, L., Niebles, J.C.: Dense-captioning events in videos. In: IEEE International Conference on Computer Vision, ICCV 2017, Venice, October 22–29, 2017, pp. 706–715 (2017)
67. Lee, P., Byun, H.: Bam-detr: boundary-aligned moment detection transformer for temporal sentence grounding in videos. arXiv preprint arXiv:2312.00083 (2023)
68. Lei, J., Yu, L., Berg, T.L., Bansal, M.: Tvr: a large-scale dataset for video-subtitle moment retrieval. In: Computer Vision—ECCV 2020: 16th European Conference, Glasgow, August 23–28, 2020, Proceedings, Part XXI 16, pp. 447–463. Springer (2020)
69. Lei, J., Berg, T.L., Bansal, M.: Detecting moments and highlights in videos via natural language queries. Adv. Neural Inf. Process. Syst. **34**, 11846–11858 (2021)
70. Li, L., Chen, Y.C., Cheng, Y., Gan, Z., Yu, L., Liu, J.: Hero: hierarchical encoder for video+language omni-representation pre-training. arXiv preprint arXiv:2005.00200 (2020)

71. Li, M., Wang, H., Zhang, W., Miao, J., Zhao, Z., Zhang, S., Ji, W., Wu, F.: Winner: weakly-supervised hierarchical decomposition and alignment for spatio-temporal video grounding. In: Proceedings of the IEEE/CVF Conference on Computer Vision and Pattern Recognition, pp. 23090–23099 (2023)
72. Lin, Z., Zhao, Z., Zhang, Z., Wang, Q., Liu, H.: Weakly-supervised video moment retrieval via semantic completion network. In: Proceedings of the AAAI Conference on Artificial Intelligence, vol. 34, pp. 11539–11546 (2020)
73. Lin, K.Q., Zhang, P., Chen, J., Pramanick, S., Gao, D., Wang, A.J., Yan, R., Shou, M.Z.: Univtg: towards unified video-language temporal grounding. In: Proceedings of the IEEE/CVF International Conference on Computer Vision, pp. 2794–2804 (2023)
74. Lin, Z., Tan, C., Hu, J.F., Jin, Z., Ye, T., Zheng, W.S.: Collaborative static and dynamic vision-language streams for spatio-temporal video grounding. In: Proceedings of the IEEE/CVF Conference on Computer Vision and Pattern Recognition, pp. 23100–23109 (2023)
75. Liu, B., Yeung, S., Chou, E., Huang, D.A., Fei-Fei, L., Niebles, J.C.: Temporal modular networks for retrieving complex compositional activities in videos. In: Proceedings of the European Conference on Computer Vision (ECCV), pp. 552–568 (2018)
76. Liu, M., Wang, X., Nie, L., He, X., Chen, B., Chua, T.: Attentive moment retrieval in videos. In: The 41st International ACM SIGIR Conference on Research & Development in Information Retrieval, SIGIR 2018, Ann Arbor, July 08–12, 2018, pp. 15–24 (2018)
77. Liu, M., Wang, X., Nie, L., Tian, Q., Chen, B., Chua, T.: Cross-modal moment localization in videos. In: 2018 ACM Multimedia Conference on Multimedia Conference, MM 2018, Seoul, October 22–26, 2018, pp. 843–851 (2018)
78. Liu, D., Qu, X., Dong, J., Zhou, P.: Reasoning step-by-step: temporal sentence localization in videos via deep rectification-modulation network. In: Proceedings of the 28th International Conference on Computational Linguistics, pp. 1841–1851 (2020)
79. Liu, D., Qu, X., Liu, X., Dong, J., Zhou, P., Xu, Z.: Jointly cross- and self-modal graph attention network for query-based moment localization. In: MM '20: The 28th ACM International Conference on Multimedia, Virtual Event, Seattle, October 12–16, 2020, pp. 4070–4078 (2020)
80. Liu, D., Qu, X., Dong, J., Zhou, P., Cheng, Y., Wei, W., Xu, Z., Xie, Y.: Context-aware biaffine localizing network for temporal sentence grounding. In: Proceedings of the IEEE/CVF Conference on Computer Vision and Pattern Recognition, pp. 11235–11244 (2021)
81. Liu, D., Qu, X., Wang, Y., Di, X., Zou, K., Cheng, Y., Xu, Z., Zhou, P.: Unsupervised temporal video grounding with deep semantic clustering. In: Proceedings of the AAAI Conference on Artificial Intelligence, vol. 36, pp. 1683–1691 (2022)
82. Liu, Y., Li, S., Wu, Y., Chen, C.W., Shan, Y., Qie, X.: Umt: unified multi-modal transformers for joint video moment retrieval and highlight detection. In: Proceedings of the IEEE/CVF Conference on Computer Vision and Pattern Recognition, pp. 3042–3051 (2022)
83. Liu, S., Zeng, Z., Ren, T., Li, F., Zhang, H., Yang, J., Li, C., Yang, J., Su, H., Zhu, J., et al.: Grounding dino: marrying dino with grounded pre-training for open-set object detection. arXiv preprint arXiv:2303.05499 (2023)
84. Lu, C., Chen, L., Tan, C., Li, X., Xiao, J.: DEBUG: a dense bottom-up grounding approach for natural language video localization. In: Proceedings of the 2019 Conference on Empirical Methods in Natural Language Processing and the 9th International Joint Conference on Natural Language Processing (EMNLP-IJCNLP), pp. 5144–5153 (2019)
85. Lu, Y., Quan, R., Zhu, L., Yang, Y.: Zero-shot video grounding with pseudo query lookup and verification. IEEE Trans. Image Process. **33**, 1643–1654 (2024)
86. Luo, J., Li, Y., Pan, Y., Yao, T., Chao, H., Mei, T.: Coco-bert: improving video-language pre-training with contrastive cross-modal matching and denoising. In: Proceedings of the 29th ACM International Conference on Multimedia, pp. 5600–5608 (2021)
87. Luo, F., Chen, S., Chen, J., Wu, Z., Jiang, Y.G.: Self-supervised learning for semi-supervised temporal language grounding. IEEE Trans. Multimedia. **25**, 7747–775 (2022)
88. Luo, D., Huang, J., Gong, S., Jin, H., Liu, Y.: Zero-shot video moment retrieval from frozen vision-language models. In: Proceedings of the IEEE/CVF Winter Conference on Applications of Computer Vision, pp. 5464–5473 (2024)

89. Lv, Z., Su, B., Wen, J.R.: Counterfactual cross-modality reasoning for weakly supervised video moment localization. In: Proceedings of the 31st ACM International Conference on Multimedia, pp. 6539–6547 (2023)
90. Ma, M., Yoon, S., Kim, J., Lee, Y., Kang, S., Yoo, C.D.: Vlanet: video-language alignment network for weakly-supervised video moment retrieval. In: European Conference on Computer Vision, pp. 156–171. Springer (2020)
91. Ma, F., Jin, X., Wang, H., Huang, J., Zhu, L., Feng, J., Yang, Y.: Temporal perceiving video-language pre-training. arXiv preprint arXiv:2301.07463 (2023)
92. Miech, A., Zhukov, D., Alayrac, J.B., Tapaswi, M., Laptev, I., Sivic, J.: Howto100m: learning a text-video embedding by watching hundred million narrated video clips. In: Proceedings of the IEEE/CVF International Conference on Computer Vision, pp. 2630–2640 (2019)
93. Mithun, N.C., Paul, S., Roy-Chowdhury, A.K.: Weakly supervised video moment retrieval from text queries. In: IEEE Conference on Computer Vision and Pattern Recognition, CVPR 2019, Long Beach, June 16–20, 2019, pp. 11592–11601 (2019)
94. Moon, W., Hyun, S., Lee, S., Heo, J.P.: Correlation-guided query-dependency calibration in video representation learning for temporal grounding. arXiv preprint arXiv:2311.08835 (2023)
95. Moon, W., Hyun, S., Park, S., Park, D., Heo, J.P.: Query-dependent video representation for moment retrieval and highlight detection. In: Proceedings of the IEEE/CVF Conference on Computer Vision and Pattern Recognition, pp. 23023–23033 (2023)
96. Mu, F., Mo, S., Li, Y.: Snag: scalable and accurate video grounding. In: Proceedings of the IEEE/CVF Conference on Computer Vision and Pattern Recognition, pp. 18930–18940 (2024)
97. Mun, J., Cho, M., Han, B.: Local-global video-text interactions for temporal grounding. In: 2020 IEEE/CVF Conference on Computer Vision and Pattern Recognition, CVPR 2020, Seattle, June 13–19, 2020, pp. 10807–10816 (2020)
98. Nagrani, A., Seo, P.H., Seybold, B., Hauth, A., Manen, S., Sun, C., Schmid, C.: Learning audio-video modalities from image captions. In: European Conference on Computer Vision, pp. 407–426. Springer (2022)
99. Nam, J., Ahn, D., Kang, D., Ha, S.J., Choi, J.: Zero-shot natural language video localization. In: Proceedings of the IEEE/CVF International Conference on Computer Vision, pp. 1470–1479 (2021)
100. Nan, G., Qiao, R., Xiao, Y., Liu, J., Leng, S., Zhang, H., Lu, W.: Interventional video grounding with dual contrastive learning. In: Proceedings of the IEEE/CVF Conference on Computer Vision and Pattern Recognition, pp. 2765–2775 (2021)
101. Otani, M., Nakashima, Y., Rahtu, E., Heikkilä, J.: Uncovering hidden challenges in query-based video moment retrieval. In: 31st British Machine Vision Conference 2020, BMVC 2020, Virtual Event, September 7–10, 2020 (2020)
102. Pan, Y., Li, Y., Luo, J., Xu, J., Yao, T., Mei, T.: Auto-captions on gif: a large-scale video-sentence dataset for vision-language pre-training. arXiv preprint arXiv:2007.02375 (2020)
103. Pan, Y., He, X., Gong, B., Lv, Y., Shen, Y., Peng, Y., Zhao, D.: Scanning only once: an end-to-end framework for fast temporal grounding in long videos. In: Proceedings of the IEEE/CVF International Conference on Computer Vision, pp. 13767–13777 (2023)
104. Qu, X., Tang, P., Zou, Z., Cheng, Y., Dong, J., Zhou, P., Xu, Z.: Fine-grained iterative attention network for temporal language localization in videos. In: MM '20: The 28th ACM International Conference on Multimedia, Virtual Event, Seattle, October 12–16, 2020, pp. 4280–4288 (2020)
105. Regneri, M., Rohrbach, M., Wetzel, D., Thater, S., Schiele, B., Pinkal, M.: Grounding action descriptions in videos. Trans. Assoc. Comput. Linguist. **1**, 25–36 (2013)
106. Rezatofighi, H., Tsoi, N., Gwak, J., Sadeghian, A., Reid, I., Savarese, S.: Generalized intersection over union: a metric and a loss for bounding box regression. In: Proceedings of the IEEE/CVF Conference on Computer Vision and Pattern Recognition, pp. 658–666 (2019)
107. Rodriguez, C., Marrese-Taylor, E., Saleh, F.S., Li, H., Gould, S.: Proposal-free temporal moment localization of a natural-language query in video using guided attention. In:

Proceedings of the IEEE/CVF Winter Conference on Applications of Computer Vision, pp. 2464–2473 (2020)
108. Rodriguez-Opazo, C., Marrese-Taylor, E., Fernando, B., Li, H., Gould, S.: Dori: discovering object relationships for moment localization of a natural language query in a video. In: Proceedings of the IEEE/CVF Winter Conference on Applications of Computer Vision, pp. 1079–1088 (2021)
109. Rohrbach, M., Regneri, M., Andriluka, M., Amin, S., Pinkal, M., Schiele, B.: Script data for attribute-based recognition of composite activities. In: European Conference on Computer Vision, pp. 144–157. Springer (2012)
110. Shi, J., Xu, J., Gong, B., Xu, C.: Not all frames are equal: weakly-supervised video grounding with contextual similarity and visual clustering losses. In: Proceedings of the IEEE/CVF Conference on Computer Vision and Pattern Recognition, pp. 10444–10452 (2019)
111. Shi, F., Huang, W., Wang, L.: End-to-end dense video grounding via parallel regression. Comput. Vis. Image Underst. **242**, 103980 (2024)
112. Sigurdsson, G.A., Varol, G., Wang, X., Farhadi, A., Laptev, I., Gupta, A.: Hollywood in homes: crowdsourcing data collection for activity understanding. In: European Conference on Computer Vision, pp. 510–526. Springer (2016)
113. Soldan, M., Xu, M., Qu, S., Tegner, J., Ghanem, B.: Vlg-net: video-language graph matching network for video grounding. In: Proceedings of the IEEE/CVF International Conference on Computer Vision, pp. 3224–3234 (2021)
114. Soldan, M., Pardo, A., Alcázar, J.L., Caba, F., Zhao, C., Giancola, S., Ghanem, B.: Mad: a scalable dataset for language grounding in videos from movie audio descriptions. In: Proceedings of the IEEE/CVF Conference on Computer Vision and Pattern Recognition, pp. 5026–5035 (2022)
115. Song, X., Han, Y.: Val: visual-attention action localizer. In: Pacific rim Conference on Multimedia, pp. 340–350. Springer (2018)
116. Song, Y., Wang, J., Ma, L., Yu, Z., Yu, J.: Weakly-supervised multi-level attentional reconstruction network for grounding textual queries in videos. ArXiv preprint abs/2003.07048 (2020)
117. Stroud, J.C., McCaffrey, R., Mihalcea, R., Deng, J., Russakovsky, O.: Compositional temporal visual grounding of natural language event descriptions. arXiv preprint arXiv:1912.02256 (2019)
118. Su, R., Yu, Q., Xu, D.: Stvgbert: a visual-linguistic transformer based framework for spatio-temporal video grounding. In: Proceedings of the IEEE/CVF International Conference on Computer Vision, pp. 1533–1542 (2021)
119. Sun, L., Wang, P., Wang, L., Sun, J., Okatani, T.: Zero-shot temporal event localisation: label-free, training-free, domain-free. IET Comput. Vis. **17**(5), 599–613 (2023)
120. Sun, H., Zhou, M., Chen, W., Xie, W.: Tr-detr: task-reciprocal transformer for joint moment retrieval and highlight detection. In: Proceedings of the AAAI Conference on Artificial Intelligence, vol. 38, pp. 4998–5007 (2024)
121. Sun, X., Shi, L., Wang, L., Zhou, S., Xia, K., Wang, Y., Hua, G.: Diversifying query: region-guided transformer for temporal sentence grounding. arXiv preprint arXiv:2406.00143 (2024)
122. Tan, R., Xu, H., Saenko, K., Plummer, B.A.: Logan: latent graph co-attention network for weakly-supervised video moment retrieval. In: Proceedings of the IEEE/CVF Winter Conference on Applications of Computer Vision, pp. 2083–2092 (2021)
123. Tan, C., Lin, Z., Hu, J.F., Zheng, W.S., Lai, J.: Hierarchical semantic correspondence networks for video paragraph grounding. In: Proceedings of the IEEE/CVF Conference on Computer Vision and Pattern Recognition, pp. 18973–18982 (2023)
124. Tan, C., Lai, J., Zheng, W.S., Hu, J.F.: Siamese learning with joint alignment and regression for weakly-supervised video paragraph grounding. In: Proceedings of the IEEE/CVF Conference on Computer Vision and Pattern Recognition, pp. 13569–13580 (2024)
125. Tang, Z., Liao, Y., Liu, S., Li, G., Jin, X., Jiang, H., Yu, Q., Xu, D.: Human-centric spatio-temporal video grounding with visual transformers. IEEE Trans. Circuits Syst. Video Technol. **32**(12), 8238–8249 (2021)

126. Vaswani, A., Shazeer, N., Parmar, N., Uszkoreit, J., Jones, L., Gomez, A.N., Kaiser, L., Polosukhin, I.: Attention is all you need. In: Advances in Neural Information Processing Systems 30: Annual Conference on Neural Information Processing Systems 2017, December 4–9, 2017, Long Beach, pp. 5998–6008 (2017)
127. Wang, W., Huang, Y., Wang, L.: Language-driven temporal activity localization: a semantic matching reinforcement learning model. In: IEEE Conference on Computer Vision and Pattern Recognition, CVPR 2019, Long Beach, June 16–20, 2019, pp. 334–343 (2019)
128. Wang, H., Zha, Z., Chen, X., Xiong, Z., Luo, J.: Dual path interaction network for video moment localization. In: MM '20: The 28th ACM International Conference on Multimedia, Virtual Event, Seattle, October 12–16, 2020, pp. 4116–4124 (2020)
129. Wang, J., Ma, L., Jiang, W.: Temporally grounding language queries in videos by contextual boundary-aware prediction. In: Proceedings of the AAAI Conference on Artificial Intelligence, vol. 34, pp. 12168–12175 (2020)
130. Wang, H., Zha, Z.J., Li, L., Liu, D., Luo, J.: Structured multi-level interaction network for video moment localization via language query. In: Proceedings of the IEEE/CVF Conference on Computer Vision and Pattern Recognition, pp. 7026–7035 (2021)
131. Wang, Y., Zhou, W., Li, H.: Fine-grained semantic alignment network for weakly supervised temporal language grounding. In: Findings of the Association for Computational Linguistics: EMNLP 2021, pp. 89–99 (2021)
132. Wang, Z., Chen, J., Jiang, Y.G.: Visual co-occurrence alignment learning for weakly-supervised video moment retrieval. In: Proceedings of the 29th ACM International Conference on Multimedia, pp. 1459–1468 (2021)
133. Wang, G., Wu, X., Liu, Z., Yan, J.: Prompt-based zero-shot video moment retrieval. In: Proceedings of the 30th ACM International Conference on Multimedia, pp. 413–421 (2022)
134. Wang, P., Sun, L., Wang, L., Sun, J.: Zero-shot video grounding for automatic video understanding in sustainable smart cities. Sustainability **15**(1), 153 (2022)
135. Wang, W., Gao, J., Xu, C.: Weakly-supervised video object grounding via causal intervention. IEEE Trans. Pattern Anal. Mach. Intell. **45**(3), 3933–3948 (2022)
136. Wang, Y., Li, K., Li, X., Yu, J., He, Y., Chen, G., Pei, B., Zheng, R., Xu, J., Wang, Z., et al.: Internvideo2: scaling video foundation models for multimodal video understanding. arXiv preprint arXiv:2403.15377 (2024)
137. Wasim, S.T., Naseer, M., Khan, S., Yang, M.H., Khan, F.S.: Videogrounding-dino: towards open-vocabulary spatio-temporal video grounding. In: Proceedings of the IEEE/CVF Conference on Computer Vision and Pattern Recognition, pp. 18909–18918 (2024)
138. Wattasseril, J.I., Shekhar, S., Döllner, J., Trapp, M.: Zero-shot video moment retrieval using blip-based models. In: International Symposium on Visual Computing, pp. 160–171. Springer (2023)
139. Wu, A., Han, Y.: Multi-modal circulant fusion for video-to-language and backward. In: Proceedings of the Twenty-Seventh International Joint Conference on Artificial Intelligence, IJCAI 2018, July 13–19, 2018, Stockholm, pp. 1029–1035 (2018)
140. Wu, J., Li, G., Han, X., Lin, L.: Reinforcement learning for weakly supervised temporal grounding of natural language in untrimmed videos. In: MM '20: The 28th ACM International Conference on Multimedia, Virtual Event, Seattle, October 12–16, 2020, pp. 1283–1291 (2020)
141. Wu, J., Li, G., Liu, S., Lin, L.: Tree-structured policy based progressive reinforcement learning for temporally language grounding in video. In: Proceedings of the AAAI Conference on Artificial Intelligence, vol. 34, pp. 12386–12393 (2020)
142. Xiao, S., Chen, L., Zhang, S., Ji, W., Shao, J., Ye, L., Xiao, J.: Boundary proposal network for two-stage natural language video localization. In: Proceedings of the AAAI Conference on Artificial Intelligence, vol. 35, pp. 2986–2994 (2021)
143. Xiao, Y., Luo, Z., Liu, Y., Ma, Y., Bian, H., Ji, Y., Yang, Y., Li, X.: Bridging the gap: a unified video comprehension framework for moment retrieval and highlight detection. In: Proceedings of the IEEE/CVF Conference on Computer Vision and Pattern Recognition, pp. 18709–18719 (2024)

144. Xu, H., He, K., Plummer, B.A., Sigal, L., Sclaroff, S., Saenko, K.: Multilevel language and vision integration for text-to-clip retrieval. In: Proceedings of the AAAI Conference on Artificial Intelligence, vol. 33, pp. 9062–9069 (2019)
145. Xu, M., Pérez-Rúa, J.M., Escorcia, V., Martinez, B., Zhu, X., Zhang, L., Ghanem, B., Xiang, T.: Boundary-sensitive pre-training for temporal localization in videos. In: Proceedings of the IEEE/CVF International Conference on Computer Vision, pp. 7220–7230 (2021)
146. Xu, M., Soldan, M., Gao, J., Liu, S., Pérez-Rúa, J.M., Ghanem, B.: Boundary-denoising for video activity localization. arXiv preprint arXiv:2304.02934 (2023)
147. Xu, Y., Sun, Y., Li, Y., Shi, Y., Zhu, X., Du, S.: Mh-detr: video moment and highlight detection with cross-modal transformer. arXiv preprint arXiv:2305.00355 (2023)
148. Xu, Y., Sun, Y., Xie, Z., Zhai, B., Du, S.: Vtg-gpt: tuning-free zero-shot video temporal grounding with gpt. Appl. Sci. **14**(5), 1894 (2024)
149. Yang, X., Feng, F., Ji, W., Wang, M., Chua, T.S.: Deconfounded video moment retrieval with causal intervention. In: Proceedings of the 44th International ACM SIGIR Conference on Research and Development in Information Retrieval, SIGIR '21, pp. 1–10 (2021)
150. Yang, A., Miech, A., Sivic, J., Laptev, I., Schmid, C.: Tubedetr: spatio-temporal video grounding with transformers. In: Proceedings of the IEEE/CVF Conference on Computer Vision and Pattern Recognition, pp. 16442–16453 (2022)
151. Yang, J., Wei, P., Li, H., Ren, Z.: Task-driven exploration: decoupling and inter-task feedback for joint moment retrieval and highlight detection. In: Proceedings of the IEEE/CVF Conference on Computer Vision and Pattern Recognition, pp. 18308–18318 (2024)
152. Yoon, S., Hong, J.W., Yoon, E., Kim, D., Kim, J., Yoon, H.S., Yoo, C.D.: Selective query-guided debiasing for video corpus moment retrieval. In: European Conference on Computer Vision, pp. 185–200. Springer (2022)
153. Yoon, S., Koo, G., Kim, D., Yoo, C.D.: Scanet: scene complexity aware network for weakly-supervised video moment retrieval. In: Proceedings of the IEEE/CVF International Conference on Computer Vision, pp. 13576–13586 (2023)
154. Yu, X., Malmir, M., He, X., Chen, J., Wang, T., Wu, Y., Liu, Y., Liu, Y.: Cross interaction network for natural language guided video moment retrieval. In: Proceedings of the 44th International ACM SIGIR Conference on Research and Development in Information Retrieval, pp. 1860–1864 (2021)
155. Yuan, Y., Ma, L., Wang, J., Liu, W., Zhu, W.: Semantic conditioned dynamic modulation for temporal sentence grounding in videos. In: Advances in Neural Information Processing Systems 32: Annual Conference on Neural Information Processing Systems 2019, NeurIPS 2019, December 8–14, 2019, Vancouver, pp. 534–544 (2019)
156. Yuan, Y., Mei, T., Zhu, W.: To find where you talk: temporal sentence localization in video with attention based location regression. In: Proceedings of the AAAI Conference on Artificial Intelligence, vol. 33, pp. 9159–9166 (2019)
157. Yuan, Y., Lan, X., Wang, X., Chen, L., Wang, Z., Zhu, W.: A closer look at temporal sentence grounding in videos: dataset and metric. In: Proceedings of the 2nd International Workshop on Human-Centric Multimedia Analysis, pp. 13–21 (2021)
158. Zeng, R., Xu, H., Huang, W., Chen, P., Tan, M., Gan, C.: Dense regression network for video grounding. In: 2020 IEEE/CVF Conference on Computer Vision and Pattern Recognition, CVPR 2020, Seattle, June 13–19, 2020, pp. 10284–10293 (2020)
159. Zeng, Y., Cao, D., Wei, X., Liu, M., Zhao, Z., Qin, Z.: Multi-modal relational graph for cross-modal video moment retrieval. In: Proceedings of the IEEE/CVF Conference on Computer Vision and Pattern Recognition, pp. 2215–2224 (2021)
160. Zhang, D., Dai, X., Wang, X., Wang, Y., Davis, L.S.: MAN: moment alignment network for natural language moment retrieval via iterative graph adjustment. In: IEEE Conference on Computer Vision and Pattern Recognition, CVPR 2019, Long Beach, June 16–20, 2019, pp. 1247–1257 (2019)

161. Zhang, S., Su, J., Luo, J.: Exploiting temporal relationships in video moment localization with natural language. In: Proceedings of the 27th ACM International Conference on Multimedia, pp. 1230–1238 (2019)
162. Zhang, Z., Lin, Z., Zhao, Z., Xiao, Z.: Cross-modal interaction networks for query-based moment retrieval in videos. In: Proceedings of the 42nd International ACM SIGIR Conference on Research and Development in Information Retrieval, SIGIR 2019, Paris, July 21–25, 2019, pp. 655–664 (2019)
163. Zhang, B., Hu, H., Lee, J., Zhao, M., Chammas, S., Jain, V., Ie, E., Sha, F.: A hierarchical multi-modal encoder for moment localization in video corpus. arXiv preprint arXiv:2011.09046 (2020)
164. Zhang, H., Sun, A., Jing, W., Zhou, J.T.: Span-based localizing network for natural language video localization. In: Proceedings of the 58th Annual Meeting of the Association for Computational Linguistics, pp. 6543–6554 (2020)
165. Zhang, S., Peng, H., Fu, J., Luo, J.: Learning 2d temporal adjacent networks for moment localization with natural language. In: Proceedings of the AAAI Conference on Artificial Intelligence, vol. 34, pp. 12870–12877 (2020)
166. Zhang, Z., Lin, Z., Zhao, Z., Zhu, J., He, X.: Regularized two-branch proposal networks for weakly-supervised moment retrieval in videos. In: MM '20: The 28th ACM International Conference on Multimedia, Virtual Event, Seattle, October 12–16, 2020, pp. 4098–4106 (2020)
167. Zhang, Z., Zhao, Z., Lin, Z., He, X., et al.: Counterfactual contrastive learning for weakly-supervised vision-language grounding. Adv. Neural Inf. Process. Syst. **33**, 18123–18134 (2020)
168. Zhang, Z., Zhao, Z., Lin, Z., Huai, B., Yuan, N.J.: Object-aware multi-branch relation networks for spatio-temporal video grounding. arXiv preprint arXiv:2008.06941 (2020)
169. Zhang, Z., Zhao, Z., Zhao, Y., Wang, Q., Liu, H., Gao, L.: Where does it exist: spatio-temporal video grounding for multi-form sentences. In: Proceedings of the IEEE/CVF Conference on Computer Vision and Pattern Recognition, pp. 10668–10677 (2020)
170. Zhang, H., Sun, A., Jing, W., Nan, G., Zhen, L., Zhou, J.T., Goh, R.S.M.: Video corpus moment retrieval with contrastive learning. In: Proceedings of the 44th International ACM SIGIR Conference on Research and Development in Information Retrieval, pp. 685–695 (2021)
171. Zhang, M., Yang, Y., Chen, X., Ji, Y., Xu, X., Li, J., Shen, H.T.: Multi-stage aggregated transformer network for temporal language localization in videos. In: Proceedings of the IEEE/CVF Conference on Computer Vision and Pattern Recognition, pp. 12669–12678 (2021)
172. Zhang, L., Radke, R.J.: Natural language video moment localization through query-controlled temporal convolution. In: Proceedings of the IEEE/CVF Winter Conference on Applications of Computer Vision, pp. 682–690 (2022)
173. Zhang, X., Zhao, P., Ji, J., Lu, X., Yin, Y.: Video corpus moment retrieval via deformable multigranularity feature fusion and adversarial training. IEEE Trans. Circuits Syst. Video Technol. **34**(8), 6686–6698 (2023)
174. Zhao, Y., Zhao, Z., Zhang, Z., Lin, Z.: Cascaded prediction network via segment tree for temporal video grounding. In: Proceedings of the IEEE/CVF Conference on Computer Vision and Pattern Recognition, pp. 4197–4206 (2021)
175. Zheng, M., Huang, Y., Chen, Q., Liu, Y.: Weakly supervised video moment localization with contrastive negative sample mining. In: Proceedings of the AAAI Conference on Artificial Intelligence, vol. 36, pp. 3517–3525 (2022)
176. Zheng, M., Huang, Y., Chen, Q., Peng, Y., Liu, Y.: Weakly supervised temporal sentence grounding with gaussian-based contrastive proposal learning. In: Proceedings of the IEEE/CVF Conference on Computer Vision and Pattern Recognition, pp. 15555–15564 (2022)

177. Zheng, M., Cai, X., Chen, Q., Peng, Y., Liu, Y.: Training-free video temporal grounding using large-scale pre-trained models. arXiv preprint arXiv:2408.16219 (2024)
178. Zhou, L., Louis, N., Corso, J.J.: Weakly-supervised video object grounding from text by loss weighting and object interaction. arXiv preprint arXiv:1805.02834 (2018)
179. Zhou, H., Zhang, C., Luo, Y., Chen, Y., Hu, C.: Embracing uncertainty: decoupling and de-bias for robust temporal grounding. In: Proceedings of the IEEE/CVF Conference on Computer Vision and Pattern Recognition, pp. 8445–8454 (2021)

Part II
Generalized Video Grounding and Trending Directions

Part II will cover our insights for *Generalized Video Grounding* and the development of *Video Grounding* in the era of large foundation models, discussing future directions such as Out-of-Distribution settings which deserve further investigations.

Chapter 3
Generalized Video Grounding

Abstract In this chapter, we would like to generalize the concepts of video grounding definition. This includes generalizing (i) the query, the query in *Video Grounding* usually refers to a natural language sentence and (ii) the reference, which refers to the untrimmed video in *Video Grounding*. Meanwhile, the query and reference can be extended to more modalities, such as language, audio, and vision, and they are not restricted to grounding one single sentence with the reference of a video. Furthermore, the query could be generalized to an event and we will demonstrate how to connect such event-based generalized grounding with other video-and-language tasks, such as video question answering.

3.1 Motivation and Problem Formulation

Generalize Text-to-Video Grounding to Cross-modal Grounding To begin with, the query of video grounding could be non-textual. Besides the text-based query, other types of queries are also accessible for grounding tasks, e.g., a piece of audio/speech, an image, or just a video can also be leveraged as a query for grounding a temporal video segment of the same semantics. Traditional temporal grounding only aims to ground a target moment in a video according to a given text query. As illustrated in Fig. 3.1, the directed edge from node A to node B signifies that the query in modality A is grounded using the reference in modality B. Basically, any connection is valid in this concept map, but it could be observed that the commonly investigated problems always utilize the modality with richer or higher density semantics as the query to ground the modality with more sparse semantic information. For example, natural language data is often being used as the query to be grounded on video (video grounding) or image (referring expression localization), and modality A is also able to connect to itself, such as video-to-video indicates the newly proposed task named video re-localization, aiming to retrieve a segment semantically corresponding to the video-form query.

Generalize Temporal to Object Grounding Second, the grounding target could be indicated by more than temporal locations. The fine-grained spatial

X. Wang et al., *Video Grounding and Its Generalization*,
https://doi.org/10.1007/978-3-031-94837-4_3

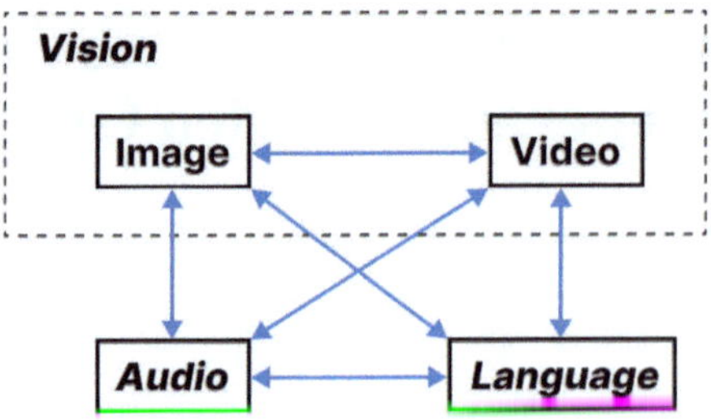

Fig. 3.1 The illustration of generalized grounding. The directed edge from A to B means using the query of modality A to ground with the reference of modality B

region (bounding box) within a frame referring to the depicted object/action can also be grounded. Thus, the grounding target becomes a continuous spatio-temporal tube (i.e., a sequence of bounding boxes) extracted from an untrimmed video via a natural language description.

Also, the temporal coordinate prediction is merely an indirect "metric" to measure whether the model truly understands the visual contents in videos. We take the same situations as examples, e.g., either the given video-related question or the generated video description refers to the interactions between different objects both temporally and spatially. The accuracy of temporal localization is not adequate since it cannot reflect the joint spatio-temporal relationships between different objects. Therefore, it is worthwhile to investigate expanding the application of restricted text-based temporal localization in the field of video grounding.

Generalize to Event Grounding As video grounding plays an important role in human-centric video understanding, e.g., given a question describing some activities in a video, the VideoQA model should first locate the relevant visual content within the video and then produce correct answer outputs according to the contextual information based on some temporal modeling and reasoning. In other words, the temporal localization process is implicitly embedded into the question-answering pipeline. Another example is dense event captioning in videos, in which the model should densely detect a series of temporal events of interest and describe them one by one in the natural language form. Therefore, such a video-to-text task can be regarded as a dual problem of video grounding, requiring the capability of semantic alignment between moment-sentence pairs as well. Here (c.f., Fig. 3.2) are some examples of video grounding stated as an intermediate/indirect task of other multi-modal problems.

Based on the concept-level generalized grounding, we believe that grounding can be used as an intermediate result to improve cross-modal semantic alignment and interpretability in other multi-modal video-centric understanding problems. In the following part, we will use several typical works as examples to show how grounding can guide multi-modal tasks with two or even three modalities (Fig. 3.3).

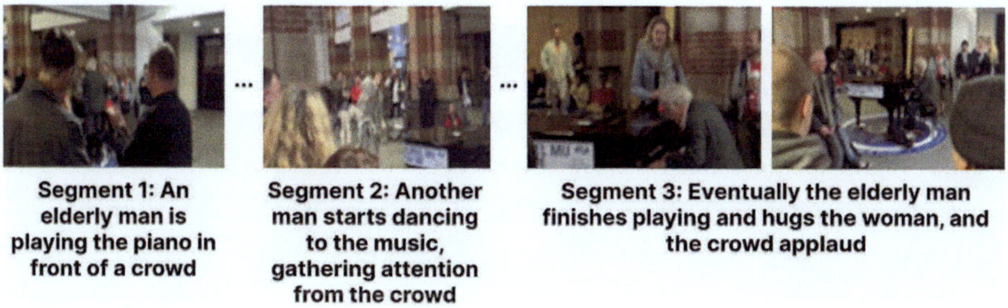

Fig. 3.2 An example of dense event captioning adopted from [10], which is a dual problem of temporal video grounding

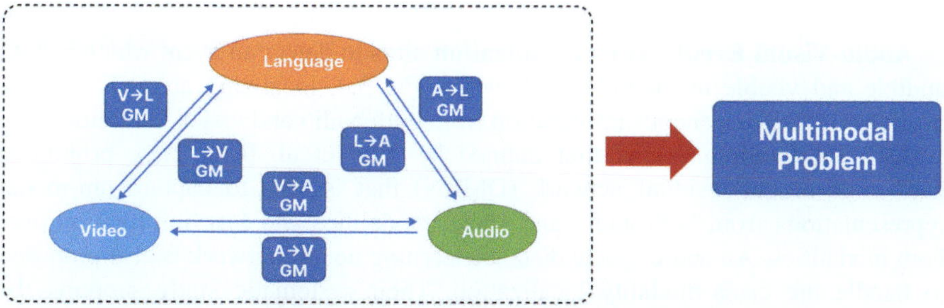

Fig. 3.3 The formulation of generalized grounding

3.2 Cross-Modal Grounding in Videos

Apart from the traditional *Video Grounding* method of applying sentence queries to ground video clips in videos, other modalities such as audio, image, and short video clips may also serve as queries for *Video Grounding*, which can be summarized as audio-visual, audio-based, image-based, and video-based temporal/spatio-temporal localization.

3.2.1 Audio-Based Temporal/Spatio-Temporal Localization

In addition to visual information, audio information attracts a lot of attention of the academic community, as videos in traditional video grounding datasets generally contain audio content, which is an additional modality of information beyond visual information in the datasets. We summarize utilizing audio information in *Video Grounding* as Audio-based Temporal/Spatio-Temporal Localization, which includes Audio-Visual Event Localization and Sound Localization.

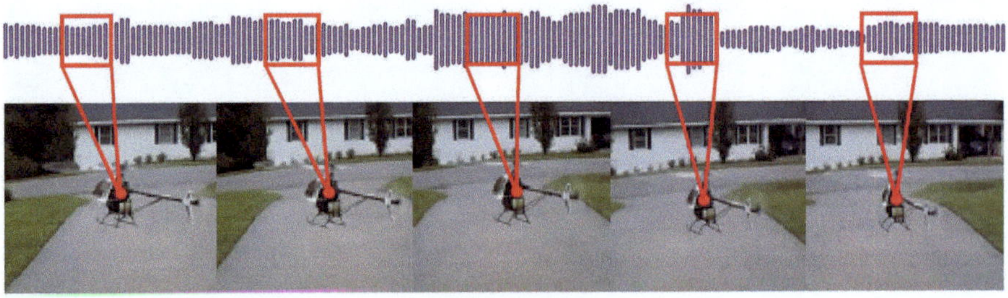

Fig. 3.4 An audio-visual event is an event both audible and visible

Audio-Visual Event (AVE) Localization aims to detect an event which is both audible and visible in videos. As shown in Fig. 3.4, localizing an AVE requires dealing with heterogeneous information from both audio and visual modalities.

AVE localization task is first defined by Tian et al. [61]. They propose a dual multi-modal residual network (DMRN) that is able to capture uni-modal representations from both audio and visual modalities and fuse information over both modalities. An audio-visual distance learning neural network is also proposed to handle the cross-modality localization. Their systematic study supports the finding that modeling jointly over auditory and visual modalities outperforms independent modeling, establishing that audio-visual localization is tractable in noisy conditions. The significance of temporal alignments for audio-visual feature fusion is further demonstrated by their ability to recognize unrelated videos and to capture semantic regions of sound sources. By exploring and learning global features of both audio and visual modalities, Lin et al. [39] propose a dual-modality sequence-sequence framework, which can be realized in either fully supervised or weakly supervised settings of AVE localization. By jointly taking both audio and visual features at each time segment as inputs, they propose a deep neural network named Audio-Visual Sequence-to-Sequence Dual Network (AVSDN) to learn global and local event information in a sequence-to-sequence manner. Comprehensive results confirm that their method performs favorably against other deep learning approaches in both fully supervised and weakly supervised settings. Inspired by the human system, which puts different focuses on specific locations, time segments, and media while performing multi-modality perception, Xuan et al. [70] propose a Cross-Modal Attention Network for temporal inconsistent audio-visual event localizationthat which is able to trade information between different modalities and achieve event localization adaptively. To better address the challenge in audio-visual event localization of fusing information from multiple modalities and obtain more robust representations, Duan et al. [11] propose a novel joint co-attention mechanism with double fusion for the audio-visual event localization task and provide qualitative visualization results shown in Fig. 3.5. Integrating double fusion leads to better representations for the AVE task by co-attending to both audio and visual modalities. In unconstrained videos, both visual and audio

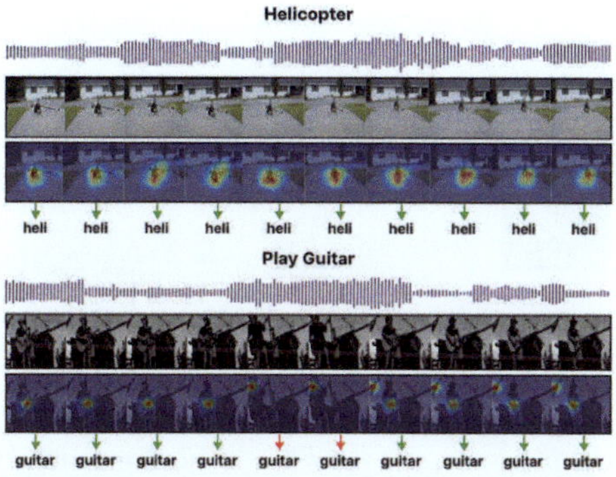

Fig. 3.5 Qualitative visualization results for the audio-visual localization task. The green arrow represents the correct prediction, and the red arrow denotes the wrong prediction. The corresponding attention maps of the original images are generated below

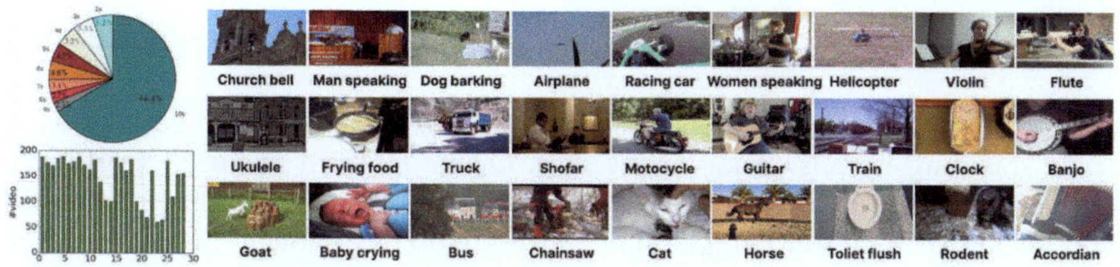

Fig. 3.6 The AVE dataset. The examples in the dataset, the distribution of videos in different categories, and the distribution of event duration are shown above

information may be inconsistent or suffer from severe background noise. To address this issue, Xia et al. [67] propose a cross-modal background suppression network for audio-visual event localization tasks, which improves localization performance by suppressing asynchronous audio-visual background frames from the examined events and reducing redundant noise.

Audio-Visual Event Dataset is a subset of AudioSet [19] constructed by Tian et al. [61]. As shown in Fig. 3.6, it contains 4143 samples of 28 event categories such as instrument performance, human daily activities, vehicle activities, and animal actions. These samples in the AVE dataset are also filled with temporal inconsistency, abrupt changes of view, and different types of noise, challenging the generalization ability of AVE models. Each sample that has been labeled with audio-visual event boundaries usually has at least one *2 seconds long* audio-visual event and is roughly *10 seconds long*.

Sound Localization aims to identify specific regions within a video that correspond to particular sounds by utilizing visual assistance. It is first proposed by Hershey et al. [24] by using a Gaussian process model to measure the mutual information from both audio and visual features. Owens and Efros [49] argue that the visual and audio components of a video could be modeled jointly through a fused multi-sensory representation. They thereby propose a multi-sensory model that learns audio-visual correspondence under self-supervised circumstances to align the audio features and visual frames, being able to localize source sound from the video. A limited number of audio sources can be handled by the early-

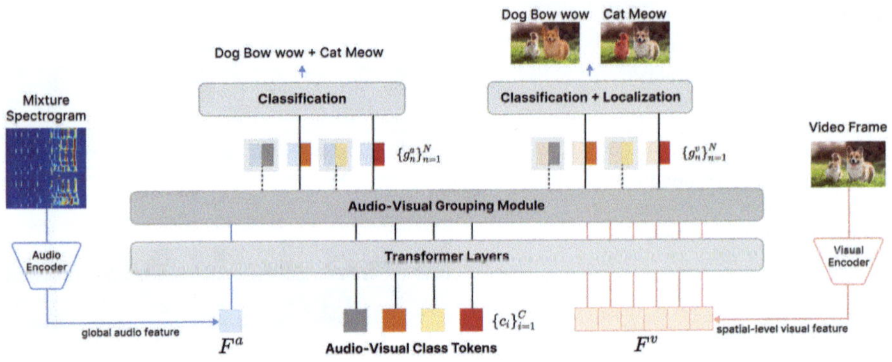

Fig. 3.7 Illustration of the AVGN

stage sound localization method, which primarily uses audio-visual correlation to locate sounding objects. Additionally, it is unable to learn compact class-aware representations for individual sources. To address these issues, Mo and Tian [47] propose a novel audio-visual grouping network (AVGN), shown in Fig. 3.7, which can directly learn category-wise semantic features for each source from the mixture of audio input and visual frame to localize multiple sources simultaneously.

Sound Localization Dataset includes MUSIC and VGGSound. MUSIC contains 448 untrimmed music videos from YouTube, ranging from 11 instrument categories of solos and duets. MUSIC-Solo is a subset of MUSIC for single-source localization, including 358 solo videos for training and 90 solo videos for evaluation. With 124 duet videos used for training and 17 for evaluation, MUSIC-Duet is used to assess multi-source localization. VGGSound is a large-scale dataset consisting of more than 200k videos for 300 audio classes. A subset of the VGGSound dataset VGGSound-Instruments is applied for sound localization, and it contains 32k video clips of 10 s lengths from 37 musical instrument classes.

3.2.2 *Image-Based Temporal/Spatio-Temporal Localization*

Image-based Activity Localization is a work using image queries to localize unseen activities in one untrimmed video. Traditional video grounding techniques are limited to a predetermined set of actions in real-world applications, and it can be challenging to identify invisible activities that are not present in the training dataset. However, these problems can be solved using an image query as a reference and localizing the unseen activities by matching visual contents to objects, appearances, motions, and interactions. This task faces many challenges, such as eliminating the influence of semantically inessential content in image queries, fuzzy localization of inaccurate image queries, and determining the precise temporal boundaries. As shown in Fig. 3.8a, searching an image query such as "playing soccer on the grass"

Fig. 3.8 Image-based activity localization. (**a**) playing soccer on the grass. (**b**) playing the violin

or "on the bench" requires precise image representation learning and a video-based attention method to highlight the crucial query information. Furthermore, as shown in Fig. 3.8b, users aim to retrieve the activity "playing the violin" but give an image query about "playing the guitar", which requires the multi-level matching of both accurate and rough semantics. Zhang et al. [80] first propose this task and employ a novel self-attention interaction localizer to retrieve unseen activities in an end-to-end setting.

ActivityIBAL is a reorganized **Image-based Activity Localization** dataset constructed from the ActivityNet dataset. It includes 160 action classes for training, 20 classes for validation, and the remaining 20 classes for testing. These videos and their classes used in validation and testing do not appear during the training process to guarantee the unseen activities' requirements for evaluation.

3.2.3 Video-Based Temporal/Spatio-Temporal Localization

As shown in Fig. 3.9, **Video re-localization (VRL)** takes a video as the input query, aiming to find a segment in reference videos that semantically corresponds to the given query video, which is an important emerging task implicating many applications, such as fast seeking in videos, video copy detection, and video surveillance. Meanwhile, it is also a challenging research task since the visual appearance of a semantic concept in videos can have huge variations. Feng et al. [15] first define the VRL task and propose a cross-gated bilinear matching model in order to capture query reference interactions. Subsequently, a wrap long short-term memory (LSTM) network is proposed to extend the temporal localization to spatio-temporal localization [16]. Chen et al. [6] propose a graph feature pyramid model with dense predictions where a frame feature pyramid is used to generate multi-level semantic features of videos, and the scene relationships among them are then encoded using graph convolutions. For the methods above, models would predict the probability of each video frame being the start or end boundary of the target moment and then obtain the final localization result. Meanwhile, Huo et al. [28] propose a semantic relevance learning framework that includes a

Fig. 3.9 Video re-localization. The top video is a video clip of an action performed by two characters. The middle video is a whole episode containing the same action happening in a different environment marked by the green rectangle. The bottom is a video containing the same action but performed by two real people. Video re-localization requires accurately retrieving the start and end times of the green segment in the middle video and the bottom video through the given top video

fine-grained feature interaction module and semantic relevance measure module to jointly consider both the spatial and temporal dimensions of videos to discover the intrinsic semantic relevance between a pair of input videos. The most pertinent proposal is chosen as the localization outcome after their reference video is first divided into several video proposals.

To be effective, the aforementioned techniques must be trained in a fully supervised environment, which restricts their scalability and practicality in real-world applications and necessitates extensive manual annotation of temporal boundaries for target moments. To address these issues, many weakly supervised methods that only require video-level annotations of query-reference pairs have also been proposed. Huang et al. [25] propose a multiscale attention approach where the attention between multiscale temporal features is used to estimate the similarity between videos. The probability of each frame that belongs to the target moment is predicted, and the longest moment with probabilities above a set threshold would be selected as the localization result. Tang et al. [59] propose an attention feature matching model to generate frame-specific query features at each reference frame and then compute the similarities between the query and reference frames. Additionally, they use a sliding window technique to produce a number of potential temporal grounding moments; the final localized moment would be the one that most closely resembles the others. Instead of using fixed feature encoding and moment localization modules, Zhou et al. [87] propose a multi-agent-reinforced switchable network, which applies two agent-based modules to adaptively guide the feature encoding and moment localization, solving the limitation of the flexibility of feature representation and the accuracy of moment localization, and performing competitively with fully supervised methods.

ActivityNet-VRL is a **video re-localization** dataset constructed by Feng et al. [15] from the ActivityNet dataset. This dataset consists of 9530 untrimmed videos from 200 action classes. The action moment in each untrimmed video can be a query video, and its paired reference video can be selected from the untrimmed videos using the same action class. The query-reference video pairs are randomly matched during training and fixed during validation and testing. Furthermore, in weakly supervised setting, the temporal location of the action moment in the reference video is not provided during the training stage.

3.3 Event Grounding for Other Multi-Modal Problems

Apart from simply grounding, which requires event queries from the video, Video Grounding also serves as an intermediate task for video content understanding, which includes Video Question Answering (VideoQA), Audio-visual Question Answering (AVQA), and Visual Reasoning.

3.3.1 Video Question Answering

Although existing VideoQA methods have achieved relatively good performance, they do not explore whether each model part is correctly operating based on the designing logic and whether each model part correctly grounds related sub-events. As shown in Fig. 3.10, we can re-formulate VideoQA as event grounding according to the question.

The definition of video question-answering tasks can be formulated as following: given an input video $V = \{v_1, \ldots, v_l\}$ with l frames and a corresponding question Q in natural language with a ground truth answer A, either directly from the question alone or from a set of candidate options $A \in A_{cands}$. The task is to develop a model M such that:

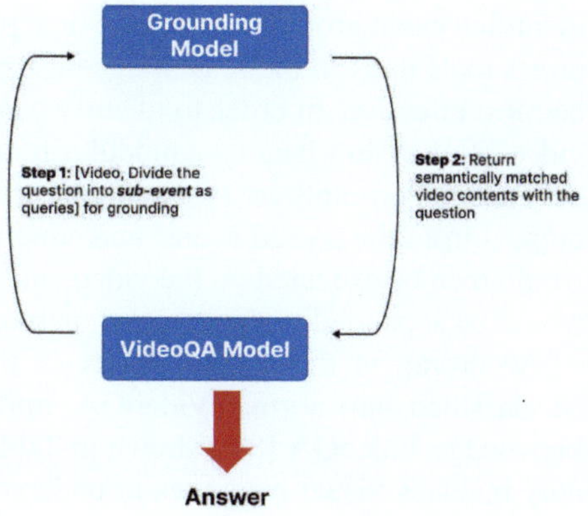

Fig. 3.10 Instantiated VideoQA task as one of the concept-level generalized grounding problems

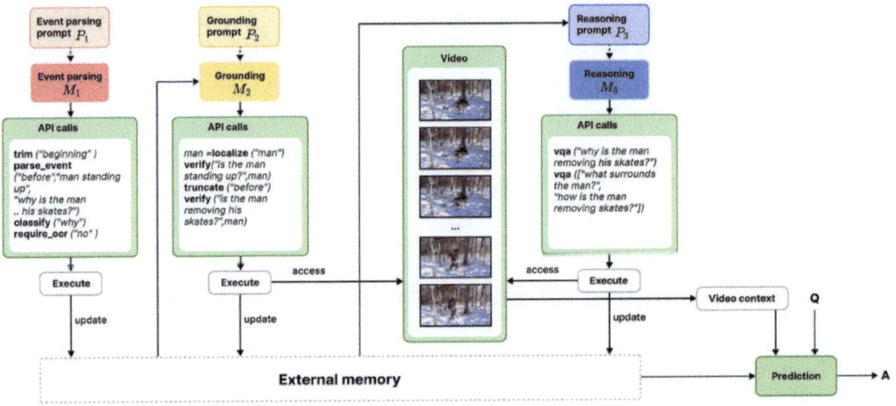

Fig. 3.11 Modular reasoning for video question answering. The grounding module plays an important role in the sub-tasks of video question answering

$$M(V, Q, [A_{cands}]) \rightarrow A, \tag{3.1}$$

where A_{cands} are present for closed-set VQA settings and not present for open-ended VQA.

Although a single-stage model promises an appealing simplicity, this design invariably results in fragile programs that fail to generate dependable outputs in practical applications. The fundamental challenge within the overall system stems from the complex task assigned to the single-stage model: it is required to produce a complete program prior to performing visual reasoning, without any reference to the actual video content.

To address this issue, as an example summarized in MoReVQA [45] shown in Fig. 3.11, the modular reasoning method of solving the video question-answering tasks could be divided into a multi-stage modular reasoning pipeline, which includes the event parsing stage, grounding stage, reasoning stage, and the final prediction stage. In the grounding stage, the focus is on the grounding-identified events. The identified event grounding is the critical process to help resolve ambiguities, and the direct tools used in the reasoning stage go to the temporal regions where they can be most effective. In order to identify candidate frames and temporal regions in the video using vision-language modules for entity detection and image-text alignment, the question prompts are typically constructed using the external memory state with outputs from the parsed events and conditions for the grounding model. The results would then be executed on the video, and the spatial and temporal grounding output would be appended to the external memory.

According to the data modality in the questions and answers, VideoQA can be classified into normal VideoQA, multi-modal VideoQA (MM VideoQA), and knowledge VideoQA [85], shown in Table 3.1. Normal Video Question Answering only includes visual resources to understand the question and to derive the correct answer, which emphasizes visual understanding of the video elements and reasoning

Table 3.1 VideoQA dataset statistics

Classes	VideoQA datasets
Normal VideoQA	VideoQA [78], MSRVTT-QA [69], MSVDQA [69], YouTube2Text-QA [82], EgoVQA [13], ActivityNet-QA [75], iVQA [71], ASRL-QA [81], Charades-SRL-QA [52], WildQA [5], TGIF-QA [29], SVQA [57], V2C-QA [14], CLEVRER [83], AGQA [22], NExT-QA [68]
MM VideoQA	MovieQA [60], MovieFIB [43], TVQA [32], LifeQA [4], How2QA [33], Env-QA [17], Pano-AVQA [76], DramaQA [9], MUSIC-AVQA [34], AVQA [72], Social-IQ [77]
Knowledge VideQA	PsTuts-VQA [84], KnowIT-VQA [18], KnowIT-X VQA [65], NEWSKVQA [23]

capabilities for their relations. The videos would usually be short and generated on social platforms. Meanwhile, MM VideoQA involves many other resources aside from visual content, such as subtitles and text plots from movies and TV shows. MM VideoQA mainly challenges the fusion of multi-modal information and the ability to understand long-story videos. Knowledge VideoQA focuses on the external knowledge distillation from explicit knowledge bases or commonsense reasoning, which always provides a global knowledge base for the whole dataset rather than giving paired knowledge for each question.

3.3.2 Audio-Visual Question Answering

Audio-visual question answering (AVQA) is a task that aims to answer questions regarding both audio and visual modalities in a given video. It requires grounding abilities shown in Fig. 3.12. The VideoQA model can incorporate the grounding model after it has been expanded to include multiple grounding modules. It helps endow the reasoning process with the strong power of temporal and spatial localization. Different from the Normal VideoQA tasks, AVQA tasks usually not only require paying attention to the visual objects, but the audible objects also possess an important role in answering the questions.

Traditional AVQA works normally adopt a two-stage strategy. First, an audio-visual model is trained to locate the sounding regions from a spatial perspective. Then, the question is used to highlight the key segments with audio-visual information from a temporal perspective. This method performs well for partial spatio-temporal reasoning in audio-visual scenarios but underperforms when the sounding visual regions highlighted by the independent first stage do not relate to the question. The second stage focuses on the visual information related to the question only at the segment level. To address this challenge, the AVQA model is expected to find the most informative audio-visual clues relevant to the given questions. Jiang et al. proposes a CLIP-powered Target-Aware Single-Stream (TASS) network [30] for AVQA using the image-text matching knowledge of the pre-trained model through

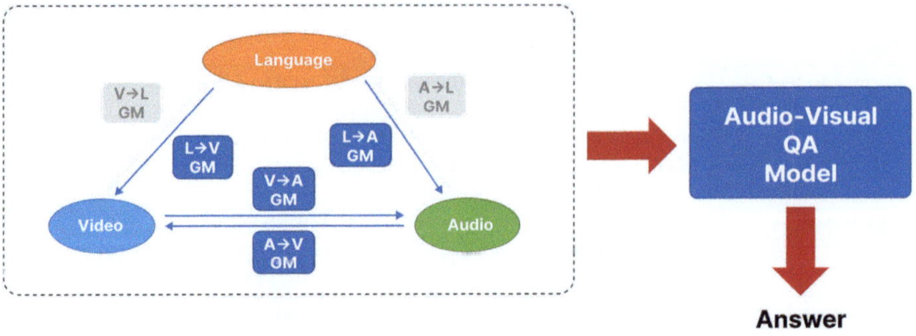

Fig. 3.12 Instantiated audio-visual QA task as another concept-level generalized grounding problem

the audio-visual matching characteristic of nature, which unifies audio-visual fusion and question-aware temporal grounding in a simplified single-stream architecture. Li et al. propose object-aware adaptive-positivity learning [37] for AVQA model optimization that adaptively selects the highly semantic-matched multi-modal pair and constrains them to have a larger similarity than the mismatched pairs.

Since the combination of audio and visual modalities is given in a video rather than an image, the audio-visual question-answering datasets are normally considered a sub-category of the video question-answering datasets (e.g., MM VideoQA). Pano-AVQA [76] studies audio-visual question-answering in panoramic video scenarios. In addition to addressing the spatiotemporal association between modalities in the AVQA task, MUSIC-AVQA [34] suggests requirements for audio-visual question answering on music performance scenes. Yang et al. propose the AVQA [72] dataset that centers around understanding various audio-visual relationships in real life.

3.3.3 Grounded Video LLMs

Incorporating Event Grounding into Video LLMs Enhancing temporal awareness of Video LLMs, which is essential for long video understanding, is also a good way to incorporate event grounding, holding the potential to improve the perceptual capabilities of Video LLMs with better temporal understanding.

With a multi-objective optimization approach that improves moment localization and language modeling, LLaViLo [42] (c.f. Fig. 3.13) uses large language models (LLMs) with adapter modules for effective multi-modal integration and temporal alignment, achieving state-of-the-art performance on benchmark datasets.

A novel Video LLM VTimeLLM [26] (c.f. Fig. 3.14) enhances fine-grained video moment understanding and reasoning through a boundary-aware three-stage training strategy; the model significantly outperforms existing Video LLMs in

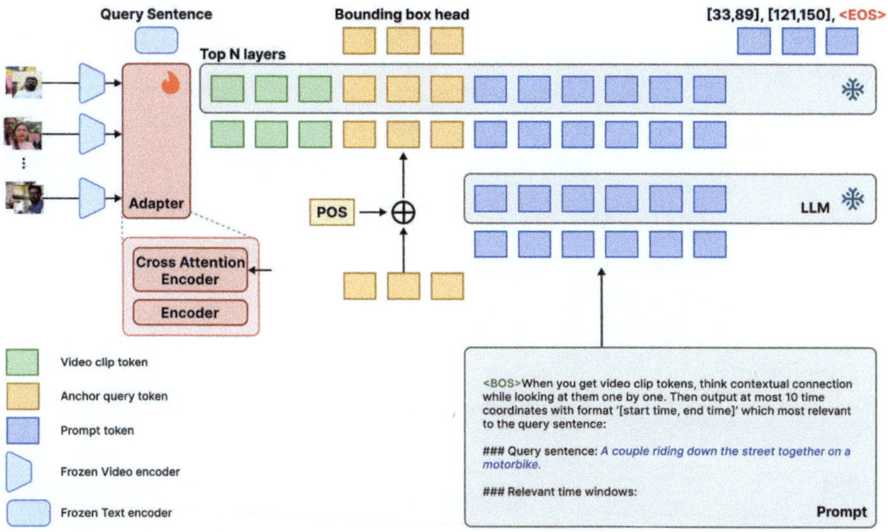

Fig. 3.13 Overview of the LLaViLo architecture

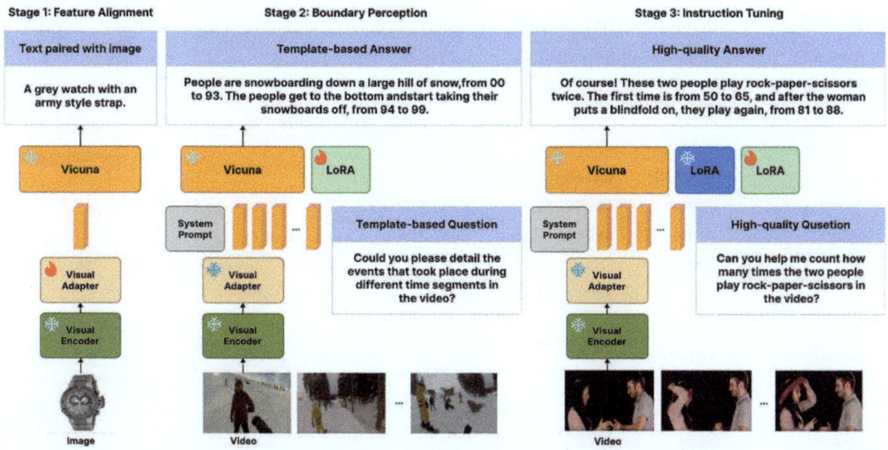

Fig. 3.14 Overview of the VTimeLLM architecture

tasks like Temporal *Video Grounding* and Dense Video Captioning, demonstrating superior cross-modal comprehension and reasoning abilities.

TimeChat [51] (c.f. Fig. 3.15) is a pioneering time-sensitive multi-modal large language model designed for a comprehensive understanding of long videos. It features a timestamp-aware frame encoder and a sliding video Q-Former, enhancing temporal localization and the ability to process videos of varying lengths. The model shows notable improvements over previous video large language models by utilizing an instruction-tuning dataset called TimeIT to enhance its performance on tasks like

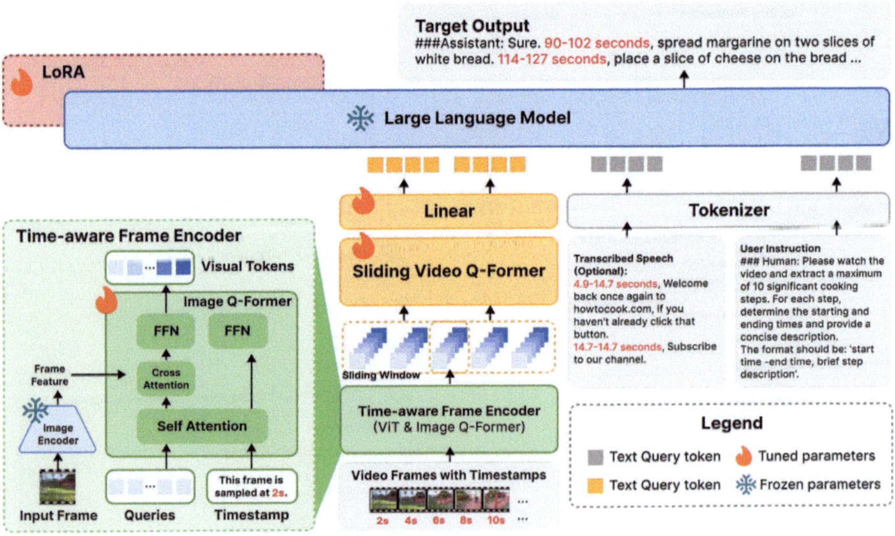

Fig. 3.15 Overview of the TimeChat architecture

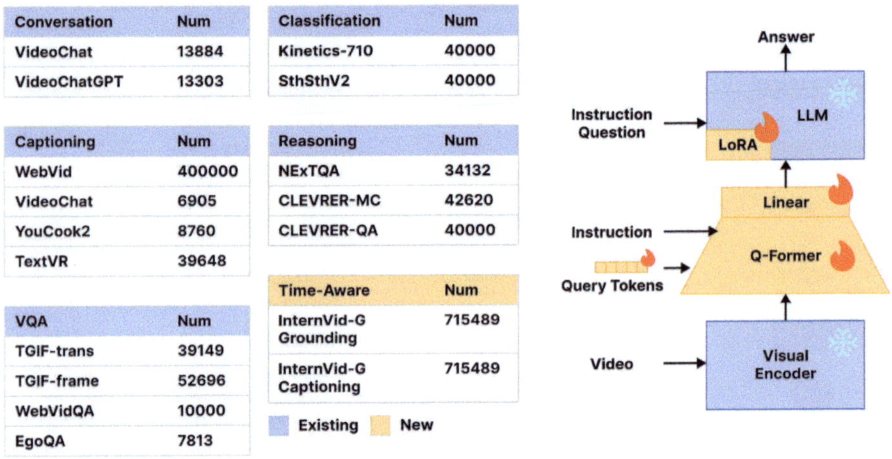

Fig. 3.16 Overview of the HawkEye training

highlight detection, temporal grounding, and dense video captioning. TimeChat's strong zero-shot capabilities and generalization to new domains position it as a versatile video assistant, holding great potential for practical applications.

HawkEye [64] (c.f. Fig. 3.16) is a pioneering video-text large language model that excels in temporal video grounding through a fully text-to-text approach. It constructs the InternVid-G dataset, which contains segment-level captions and

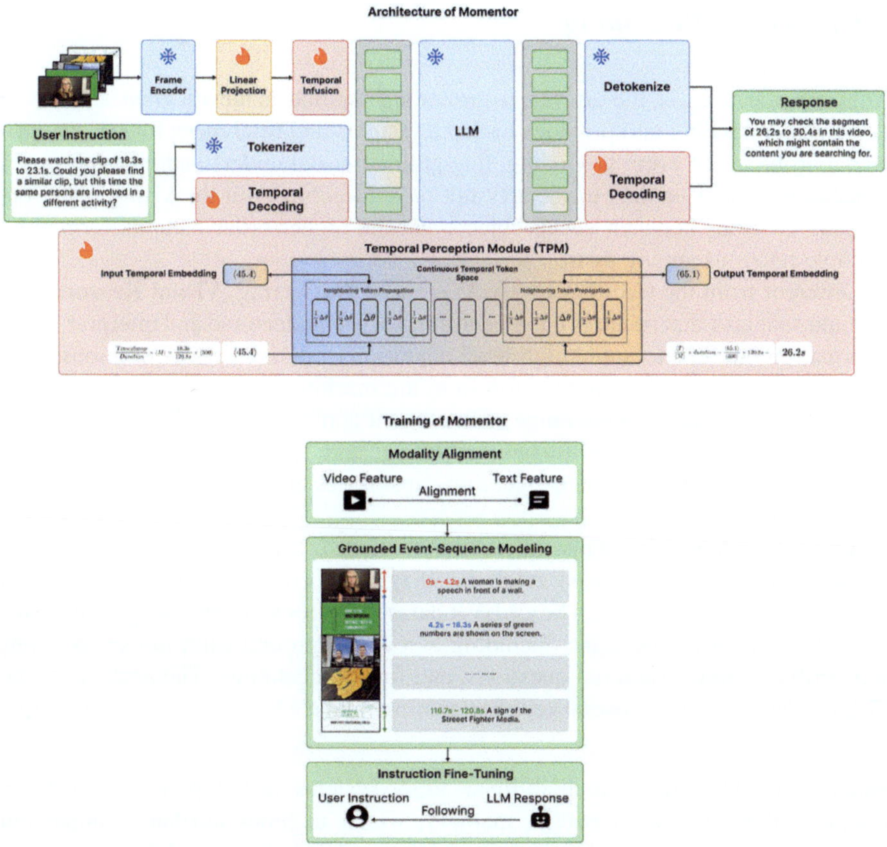

Fig. 3.17 The architecture of Momentor and training paradigm

negative spans, to enhance training for temporal grounding. Additionally, HawkEye employs a coarse-grained method for representing video segments and a recursive grounding technique, demonstrating superior performance in understanding and localizing events within videos.

Momentor [50] makes significant advancements in model architecture (c.f., Fig. 3.17), training paradigms, and fine-tuning data. First, it introduces the Temporal Perception Module (TPM), which equips the model with fine-grained positional information for temporal localization. To better train the TPM, Momentor designs a Grounded Event-Sequence Modeling approach. This method involves an additional pre-training phase using the dense video captioning task, which helps the model adapt more effectively to segment-level instruction tuning during fine tuning. Furthermore, Momentor constructs a 10M temporal dataset to enhance its ability in segment-level temporal reasoning and video comprehension.

3.3.4 *Visual Reasoning*

Although the state-of-the-art visual grounding method achieves great success in visual understanding and provides in-depth insights and references for other multi-modal problems of event grounding, fine-grained visual understanding requires the capacity of visual models to identify not only the objects in the visual scene but also the relations between visual objects and the higher-order patterns formed by multiple relations among them.

Different from the traditional Visual Question Answering, **Visual Reasoning** is an important task that requires the models not only to understand and interpret visual information but also to apply high-level cognition to derive logical solutions. This task has received significant attention from the machine learning community since its potential to enable a wide range of intelligent applications, including intelligent tutoring systems, automated image captioning, and virtual assistants. It requires models to possess both visual perception capabilities and strong logical reasoning abilities (Fig. 3.18).

To meet these requirements, CLEVR [31] is one of the typical visual reasoning datasets proposed by Johnson et al. for diagnostic compositional language and elementary visual reasoning. The CLEVR dataset applies synthetic images and automatically generates questions, including associated ground-truth object locations and attributes, tree-structured questions, and function catalogs. The analysis of the CLEVR dataset reveals several key shortcomings and challenges of traditional VQA systems:

Short-Term Memory The traditional VQA systems tested perform poorly in situations that require short-term memory, which includes attribute comparison and integer equality questions, same-attribute relationships, and tree-structured

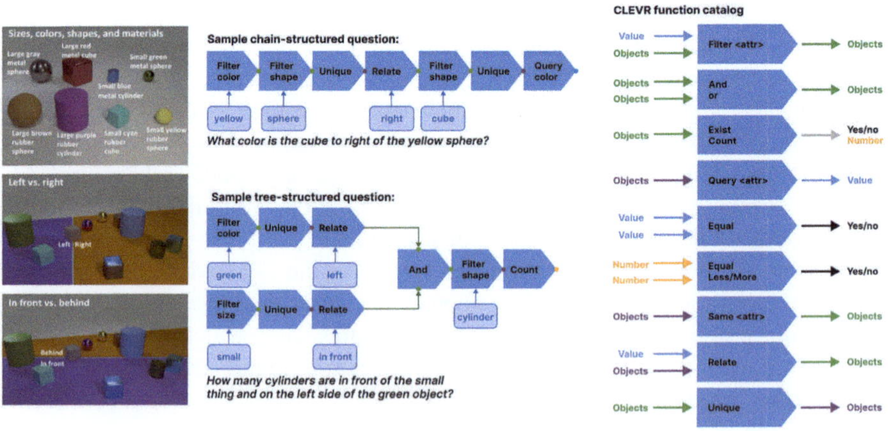

Fig. 3.18 Visualization of CLEVR, a typical visual reasoning dataset

questions. Models are often able to identify attributes of objects but struggle to compare them.

Long Reasoning Chains Traditional VQA systems struggle to answer questions requiring long chains of reasoning that are nontrivial, which include questions with large effective sizes and count and existence questions with many spatial relationships.

Spatial Relationships VQA models rely on absolute image position instead of learning the true semantics of spatial relationships.

Disentangled Representations Through training and testing VQA models on different data distributions, the models do not learn representations that properly disentangle object attributes since they learn strong biases from the training data and cannot overcome these biases when conditions change.

To address these issues, recent advanced large pre-trained models are widely applied, showing that vision-language models (VLMs) can achieve impressive performance on visual reasoning tasks even under zero-shot settings [1, 35]. Meanwhile, Large Language Models have also demonstrated robust zero-shot commonsense reasoning abilities in natural language processing applications. Several studies have attempted to combine complementary VLMs and LLMs for visual reasoning. PICa utilizes image captioning models to generate textual prompts for GPT-3 and adapts GPT-3 to solve the VQA tasks in an in-context few-shot learning manner [73]. Socratic Models allow VLMs and LLMs to communicate through prompt engineering to unlock zero-shot multi-modal reasoning capabilities [79]. Inspired by the findings in CICERO [12] that LLMs capture strong strategic planning and negotiation abilities in coordinating multiple agents, Chen et al. propose Cola [7], a novel model ensemble approach that utilizes an LLM as the coordinator in between multiple VLMs.

3.3.5 Cross-Modal Datasets for Event Grounding

With the rapid development of generalized large language models, many multi-modal large language models (MLLMs) are proposed to universally solve cross-modal problems under event grounding. In this section, we will review the literature on the datasets enabling MLLMs' capabilities for solving generalized grounding tasks. Based on the different data types, the dataset could be divided into three categories: caption, conversation, and reasoning. In addition, a large number of multi-modal large models decide to gather the previously mentioned kinds of data for integration and create their own datasets, which we also refer to as the integration dataset (Table 3.2).

Caption-based Datasets aim to build basic visual and temporal description capabilities for MLLMS. Commonly used text-to-image datasets include SBU Captions [48], MSCOCO [38], Conceptual Captions (CC-3M) [55] and LAION [53].

Table 3.2 Common cross-modal dataset statistics

Dataset type	Modalities	Datasets
Captions	Text-Image	SBU Captions [48], MSCOCO [38], CC-3M [55], LAION [53], MINT-1T [2]
	Text-Video	WebVid [3], InternVid [63], HD-VG-130M [62], YouCook2 [86], TextVR [66]
Conversation	Text-Image	VQAv2 [20], GQA [27], OK-VQA [44], AOK-VQA [54], OCR-VQA [46], TextVQA [56]
	Text-Video	TGIF-QA [29], WebVidQA [71], EgoQA [21]
Reasoning	Text-Image	CLEVR [31], VisualMRC [58]
	Text-Video	NExT-QA [68], CLEVRER [74]
Integration	Text-Image	LLaVA-Instruct [41]
	Text-Video&Image	Video-LLaVA [40], VideoChat2 [36], VideoLLaMa2 [8]

The sizes of these datasets range from 328K to 5B. Recently, MINT-1T [2] is proposed to comprise one trillion text tokens and three billion images, a 10× scale-up from existing open-source datasets, which includes previously untapped sources such as PDFs and ArXiv papers. Text-to-video datasets include WebVid [3], HD-VG-130M [62], InterVid [63], TextVR [66] and YouCook2 [86]. To comprehend the fundamental visual, aural, and temporal inputs, the caption-based dataset primarily helps to align the visual and aural modalities with the multi-modal large language model. **Conversation Datasets** aim at enhancing the capabilities of MLLMs for single-turn and multi-turn conversations when asking questions about the input image or video. A diverse set of questions would be asked about the visual content of the image and the video, including the object types, objects' number, objects' actions, objects' locations, event moment, event duration, and relative positions between objects. It is obvious that many Visual Question Answering datasets could be directly constructed as conversation datasets for MLLM training with simple formatting reorganization. These include basic VQA (VQAv2 [20], GQA [27]), knowledge-based VQA (OK-VQA [44], AOK-VQA [54]), OCR-based VQA (OCR-VQA [46], TextVQA [56]), and VideoQA (TGIF-QA [29], WebVidQA [71], and egocentric VQA from Ego4D [21]), which can not only improve the visual question answer capabilities for MLLMs in conversation, but also help the models to learn more visual, audio, and temporal knowledge.

Reasoning Datasets focus on enhancing MLLMs for diverse reasoning capacities, covering the insufficiency of the above two types of datasets, mainly focusing on the visual content itself but lacking in-depth reasoning questions. Normally, reasoning datasets require a step-by-step reasoning process by following rigorous logic. These include spatial reasoning (CLEVR [31]), reading comprehension (VisualMRC [58]), temporal reasoning (NExt-QA [68]), and spatio-temporal reasoning (CLEVRER [74]).

Due to the strong generalizing ability of large multi-modal language models, their training data usually is not limited to only one task of caption, conversation, or reasoning, but requires both simple and complex visual modal tasks. Therefore, many large multi-modal model works often do not use one single visual-audio task dataset but select subsets of several datasets from each category mentioned above for integration and adjustment, forming instruction training data that employs both image and video data for different visual-audio modal understanding tasks. For visual instruction tuning, LLaVA [41] is the first MLLM method to leverage text-only GPT4 to expand the existing bounding box and caption dataset, such as MSCOCO, to multi-modal instruction following data. In addition, Liu et al. propose LLaVA-Instruct to build on a subset of the CC-3M dataset, which contains 58k in conversations, 23k in detailed descriptions, and 77k in complex reasoning [41]. Following the development of visual instruction tuning, many video language models such as Video-LLaVA [40], VideoChat [36], and VideoLLaMa2 [8] are proposed utilizing the combination of captioning, conversation, and reasoning datasets under both text-image and text-video modalities.

References

1. Alayrac, J.B., Donahue, J., Luc, P., Miech, A., Barr, I., Hasson, Y., Lenc, K., Mensch, A., Millican, K., Reynolds, M., et al.: Flamingo: a visual language model for few-shot learning. Adv. Neural Inf. Process. Syst. **35**, 23716–23736 (2022)
2. Awadalla, A., Xue, L., Lo, O., Shu, M., Lee, H., Guha, E.K., Jordan, M., Shen, S., Awadalla, M., Savarese, S., et al.: Mint-1t: scaling open-source multimodal data by 10x: a multimodal dataset with one trillion tokens. arXiv preprint arXiv:2406.11271 (2024)
3. Bain, M., Nagrani, A., Varol, G., Zisserman, A.: Frozen in time: a joint video and image encoder for end-to-end retrieval. In: Proceedings of the IEEE/CVF International Conference on Computer Vision, pp. 1728–1738 (2021)
4. Castro, S., Azab, M., Stroud, J., Noujaim, C., Wang, R., Deng, J., Mihalcea, R.: Lifeqa: a real-life dataset for video question answering. In: Proceedings of the Twelfth Language Resources and Evaluation Conference, pp. 4352–4358 (2020)
5. Castro, S., Deng, N., Huang, P., Burzo, M., Mihalcea, R.: In-the-wild video question answering. In: Proceedings of the 29th International Conference on Computational Linguistics, pp. 5613–5635 (2022)
6. Chen, L., Lu, C., Tang, S., Xiao, J., Zhang, D., Tan, C., Li, X.: Rethinking the bottom-up framework for query-based video localization. In: Proceedings of the AAAI Conference on Artificial Intelligence, vol. 34, pp. 10551–10558 (2020)
7. Chen, L., Li, B., Shen, S., Yang, J., Li, C., Keutzer, K., Darrell, T., Liu, Z.: Large language models are visual reasoning coordinators. Adv. Neural Inf. Process. Syst. **36**, 70115–70140 (2024)
8. Cheng, Z., Leng, S., Zhang, H., Xin, Y., Li, X., Chen, G., Zhu, Y., Zhang, W., Luo, Z., Zhao, D., Bing, L.: Videollama 2: advancing spatial-temporal modeling and audio understanding in video-llms. arXiv preprint arXiv:2406.07476 (2024). https://arxiv.org/abs/2406.07476
9. Choi, S., On, K.W., Heo, Y.J., Seo, A., Jang, Y., Lee, M., Zhang, B.T.: Dramaqa: character-centered video story understanding with hierarchical qa. In: Proceedings of the AAAI Conference on Artificial Intelligence, vol. 35, pp. 1166–1174 (2021)
10. Duan, X., Huang, W., Gan, C., Wang, J., Zhu, W., Huang, J.: Weakly supervised dense event captioning in videos. In: Advances in Neural Information Processing Systems 31: Annual Conference on Neural Information Processing Systems 2018, NeurIPS 2018, December 3–8, 2018, Montréal, pp. 3063–3073 (2018)

11. Duan, B., Tang, H., Wang, W., Zong, Z., Yang, G., Yan, Y.: Audio-visual event localization via recursive fusion by joint co-attention. In: Proceedings of the IEEE/CVF Winter Conference on Applications of Computer Vision, pp. 4013–4022 (2021)
12. (FAIR)†, M.F.A.R.D.T., Bakhtin, A., Brown, N., Dinan, E., Farina, G., Flaherty, C., Fried, D., Goff, A., Gray, J., Hu, H., et al.: Human-level play in the game of diplomacy by combining language models with strategic reasoning. Science **378**(6624), 1067–1074 (2022)
13. Fan, C.: Egovqa-an egocentric video question answering benchmark dataset. In: Proceedings of the IEEE/CVF International Conference on Computer Vision Workshops, pp. 0–0 (2019)
14. Fang, Z., Gokhale, T., Banerjee, P., Baral, C., Yang, Y.: Video2commonsense: generating commonsense descriptions to enrich video captioning. arXiv preprint arXiv:2003.05162 (2020)
15. Feng, Y., Ma, L., Liu, W., Zhang, T., Luo, J.: Video re-localization. In: Proceedings of the European Conference on Computer Vision (ECCV), pp. 51–66 (2018)
16. Feng, Y., Ma, L., Liu, W., Luo, J.: Spatio-temporal video re-localization by warp lstm. In: Proceedings of the IEEE/CVF Conference on Computer Vision and Pattern Recognition, pp. 1288–1297 (2019)
17. Gao, D., Wang, R., Bai, Z., Chen, X.: Env-qa: a video question answering benchmark for comprehensive understanding of dynamic environments. In: Proceedings of the IEEE/CVF International Conference on Computer Vision, pp. 1675–1685 (2021)
18. Garcia, N., Otani, M., Chu, C., Nakashima, Y.: Knowit vqa: answering knowledge-based questions about videos. In: Proceedings of the AAAI Conference on Artificial Intelligence, vol. 34, pp. 10826–10834 (2020)
19. Gemmeke, J.F., Ellis, D.P., Freedman, D., Jansen, A., Lawrence, W., Moore, R.C., Plakal, M., Ritter, M.: Audio set: an ontology and human-labeled dataset for audio events. In: 2017 IEEE International Conference on Acoustics, Speech and Signal Processing (ICASSP), pp. 776–780. IEEE (2017)
20. Goyal, Y., Khot, T., Summers-Stay, D., Batra, D., Parikh, D.: Making the v in vqa matter: elevating the role of image understanding in visual question answering. In: Proceedings of the IEEE Conference on Computer Vision and Pattern Recognition, pp. 6904–6913 (2017)
21. Grauman, K., Westbury, A., Byrne, E., Chavis, Z., Furnari, A., Girdhar, R., Hamburger, J., Jiang, H., Liu, M., Liu, X., et al.: Ego4d: around the world in 3,000 hours of egocentric video. In: Proceedings of the IEEE/CVF Conference on Computer Vision and Pattern Recognition, pp. 18995–19012 (2022)
22. Grunde-McLaughlin, M., Krishna, R., Agrawala, M.: Agqa: a benchmark for compositional spatio-temporal reasoning. In: Proceedings of the IEEE/CVF Conference on Computer Vision and Pattern Recognition, pp. 11287–11297 (2021)
23. Gupta, P., Gupta, M.: Newskvqa: knowledge-aware news video question answering. In: Pacific-Asia Conference on Knowledge Discovery and Data Mining, pp. 3–15. Springer (2022)
24. Hershey, J., Movellan, J.: Audio vision: using audio-visual synchrony to locate sounds. Adv. Neural Inf. Process. Syst. **12**, 813–819 (1999)
25. Huang, Y.H., Hsu, K.J., Jeng, S.K., Lin, Y.Y.: Weakly-supervised video re-localization with multiscale attention model. In: Proceedings of the AAAI Conference on Artificial Intelligence, vol. 34, pp. 11077–11084 (2020)
26. Huang, B., Wang, X., Chen, H., Song, Z., Zhu, W.: Vtimellm: empower llm to grasp video moments. In: Proceedings of the IEEE/CVF Conference on Computer Vision and Pattern Recognition, pp. 14271–14280 (2024)
27. Hudson, D.A., Manning, C.D.: Gqa: a new dataset for real-world visual reasoning and compositional question answering. In: Proceedings of the IEEE/CVF Conference on Computer Vision and Pattern Recognition, pp. 6700–6709 (2019)
28. Huo, S., Zhou, Y., Wang, R., Xiang, W., Kung, S.Y.: Semantic relevance learning for video-query based video moment retrieval. IEEE Trans. Multimedia **25**, 9290–9301 (2023)
29. Jang, Y., Song, Y., Yu, Y., Kim, Y., Kim, G.: Tgif-qa: toward spatio-temporal reasoning in visual question answering. In: Proceedings of the IEEE Conference on Computer Vision and Pattern Recognition, pp. 2758–2766 (2017)

30. Jiang, Y., Yin, J.: Clip-powered tass: target-aware single-stream network for audio-visual question answering. arXiv preprint arXiv:2405.07451 (2024)
31. Johnson, J., Hariharan, B., Van Der Maaten, L., Fei-Fei, L., Lawrence Zitnick, C., Girshick, R.: Clevr: a diagnostic dataset for compositional language and elementary visual reasoning. In: Proceedings of the IEEE Conference on Computer Vision and Pattern Recognition, pp. 2901–2910 (2017)
32. Lei, J., Yu, L., Bansal, M., Berg, T.L.: Tvqa: localized, compositional video question answering. arXiv preprint arXiv:1809.01696 (2018)
33. Li, L., Chen, Y.C., Cheng, Y., Gan, Z., Yu, L., Liu, J.: Hero: hierarchical encoder for video+ language omni-representation pre-training. In: Proceedings of the 2020 Conference on Empirical Methods in Natural Language Processing (EMNLP). Association for Computational Linguistics (2020)
34. Li, G., Wei, Y., Tian, Y., Xu, C., Wen, J.R., Hu, D.: Learning to answer questions in dynamic audio-visual scenarios. In: Proceedings of the IEEE/CVF Conference on Computer Vision and Pattern Recognition, pp. 19108–19118 (2022)
35. Li, B., Zhang, Y., Chen, L., Wang, J., Pu, F., Yang, J., Li, C., Liu, Z.: Mimic-it: multi-modal in-context instruction tuning. arXiv preprint arXiv:2306.05425 (2023)
36. Li, K., Wang, Y., He, Y., Li, Y., Wang, Y., Liu, Y., Wang, Z., Xu, J., Chen, G., Luo, P., et al.: Mvbench: a comprehensive multi-modal video understanding benchmark. In: Proceedings of the IEEE/CVF Conference on Computer Vision and Pattern Recognition, pp. 22195–22206 (2024)
37. Li, Z., Guo, D., Zhou, J., Zhang, J., Wang, M.: Object-aware adaptive-positivity learning for audio-visual question answering. In: Proceedings of the AAAI Conference on Artificial Intelligence, vol. 38, pp. 3306–3314 (2024)
38. Lin, T.Y., Maire, M., Belongie, S., Hays, J., Perona, P., Ramanan, D., Dollár, P., Zitnick, C.L.: Microsoft coco: common objects in context. In: Computer Vision—ECCV 2014: 13th European Conference, Zurich, September 6–12, 2014, Proceedings, Part V 13, pp. 740–755. Springer (2014)
39. Lin, Y.B., Li, Y.J., Wang, Y.C.F.: Dual-modality seq2seq network for audio-visual event localization. In: ICASSP 2019–2019 IEEE International Conference on Acoustics, Speech and Signal Processing (ICASSP), pp. 2002–2006. IEEE (2019)
40. Lin, B., Zhu, B., Ye, Y., Ning, M., Jin, P., Yuan, L.: Video-llava: learning united visual representation by alignment before projection. arXiv preprint arXiv:2311.10122 (2023)
41. Liu, H., Li, C., Wu, Q., Lee, Y.J.: Visual instruction tuning. Adv. Neural Inf. Process. Syst. **36**, 34892–34916 (2024)
42. Ma, K., Zang, X., Feng, Z., Fang, H., Ban, C., Wei, Y., He, Z., Li, Y., Sun, H.: Llavilo: boosting video moment retrieval via adapter-based multimodal modeling. In: Proceedings of the IEEE/CVF International Conference on Computer Vision, pp. 2798–2803 (2023)
43. Maharaj, T., Ballas, N., Rohrbach, A., Courville, A., Pal, C.: A dataset and exploration of models for understanding video data through fill-in-the-blank question-answering. In: Proceedings of the IEEE Conference on Computer Vision and Pattern Recognition, pp. 6884–6893 (2017)
44. Marino, K., Rastegari, M., Farhadi, A., Mottaghi, R.: Ok-vqa: a visual question answering benchmark requiring external knowledge. In: Proceedings of the IEEE/cvf Conference on Computer Vision and Pattern Recognition, pp. 3195–3204 (2019)
45. Min, J., Buch, S., Nagrani, A., Cho, M., Schmid, C.: Morevqa: exploring modular reasoning models for video question answering. In: Proceedings of the IEEE/CVF Conference on Computer Vision and Pattern Recognition, pp. 13235–13245 (2024)
46. Mishra, A., Shekhar, S., Singh, A.K., Chakraborty, A.: Ocr-vqa: visual question answering by reading text in images. In: 2019 International Conference on Document Analysis and Recognition (ICDAR), pp. 947–952. IEEE (2019)
47. Mo, S., Tian, Y.: Audio-visual grouping network for sound localization from mixtures. In: Proceedings of the IEEE/CVF Conference on Computer Vision and Pattern Recognition (CVPR), pp. 10565–10574 (2023)

48. Ordonez, V., Kulkarni, G., Berg, T.: Im2text: describing images using 1 million captioned photographs. Adv. Neural Inf. Process. Syst. **24**, 1143–1151 (2011)
49. Owens, A., Efros, A.A.: Audio-visual scene analysis with self-supervised multisensory features. In: Proceedings of the European Conference on Computer Vision (ECCV) (2018)
50. Qian, L., Li, J., Wu, Y., Ye, Y., Fei, H., Chua, T.S., Zhuang, Y., Tang, S.: Momentor: advancing video large language model with fine-grained temporal reasoning. In: Proceedings of the 41st International Conference on Machine Learning, pp. 41340–41356 (2024)
51. Ren, S., Yao, L., Li, S., Sun, X., Hou, L.: Timechat: a time-sensitive multimodal large language model for long video understanding. In: Proceedings of the IEEE/CVF Conference on Computer Vision and Pattern Recognition, pp. 14313–14323 (2024)
52. Sadhu, A., Chen, K., Nevatia, R.: Video question answering with phrases via semantic roles. In: Proceedings of the 2021 Conference of the North American Chapter of the Association for Computational Linguistics: Human Language Technologies, pp. 2460–2478 (2021)
53. Schuhmann, C., Beaumont, R., Vencu, R., Gordon, C., Wightman, R., Cherti, M., Coombes, T., Katta, A., Mullis, C., Wortsman, M., et al.: Laion-5b: an open large-scale dataset for training next generation image-text models. Adv. Neural Inf. Process. Syst. **35**, 25278–25294 (2022)
54. Schwenk, D., Khandelwal, A., Clark, C., Marino, K., Mottaghi, R.: A-okvqa: a benchmark for visual question answering using world knowledge. In: European Conference on Computer Vision, pp. 146–162. Springer (2022)
55. Sharma, P., Ding, N., Goodman, S., Soricut, R.: Conceptual captions: a cleaned, hypernymed, image alt-text dataset for automatic image captioning. In: Proceedings of the 56th Annual Meeting of the Association for Computational Linguistics (Volume 1: Long Papers), pp. 2556–2565 (2018)
56. Singh, A., Natarajan, V., Shah, M., Jiang, Y., Chen, X., Batra, D., Parikh, D., Rohrbach, M.: Towards vqa models that can read. In: Proceedings of the IEEE/CVF Conference on Computer Vision and Pattern Recognition, pp. 8317–8326 (2019)
57. Song, X., Shi, Y., Chen, X., Han, Y.: Explore multi-step reasoning in video question answering. In: Proceedings of the 26th ACM International Conference on Multimedia, pp. 239–247 (2018)
58. Tanaka, R., Nishida, K., Yoshida, S.: Visualmrc: machine reading comprehension on document images. In: Proceedings of the AAAI Conference on Artificial Intelligence, vol. 35, pp. 13878–13888 (2021)
59. Tang, H., Zhu, J., Gao, Z., Zhuo, T., Cheng, Z.: Attention feature matching for weakly-supervised video relocalization. In: Proceedings of the 2nd ACM International Conference on Multimedia in Asia, pp. 1–7 (2021)
60. Tapaswi, M., Zhu, Y., Stiefelhagen, R., Torralba, A., Urtasun, R., Fidler, S.: Movieqa: understanding stories in movies through question-answering. In: Proceedings of the IEEE Conference on Computer Vision and Pattern Recognition, pp. 4631–4640 (2016)
61. Tian, Y., Shi, J., Li, B., Duan, Z., Xu, C.: Audio-visual event localization in unconstrained videos. In: Proceedings of the European Conference on Computer Vision (ECCV), pp. 247–263 (2018)
62. Wang, W., Yang, H., Tuo, Z., He, H., Zhu, J., Fu, J., Liu, J.: Videofactory: swap attention in spatiotemporal diffusions for text-to-video generation. In: 5th International Conference on Learning Representations. arXiv preprint arXiv:2305.10874 (2023)
63. Wang, Y., He, Y., Li, Y., Li, K., Yu, J., Ma, X., Li, X., Chen, G., Chen, X., Wang, Y., et al.: Internvid: a large-scale video-text dataset for multimodal understanding and generation. In: The Twelfth International Conference on Learning Representations (2024)
64. Wang, Y., Meng, X., Liang, J., Wang, Y., Liu, Q., Zhao, D.: Hawkeye: training video-text llms for grounding text in videos. arXiv preprint arXiv:2403.10228 (2024)
65. Wu, T., Garcia, N., Otani, M., Chu, C., Nakashima, Y., Takemura, H.: Transferring domain-agnostic knowledge in video question answering. arXiv preprint arXiv:2110.13395 (2021)
66. Wu, W., Zhao, Y., Li, Z., Li, J., Zhou, H., Shou, M.Z., Bai, X.: A large cross-modal video retrieval dataset with reading comprehension. Pattern Recognit. **157**, 110818 (2025)

67. Xia, Y., Zhao, Z.: Cross-modal background suppression for audio-visual event localization. In: Proceedings of the IEEE/CVF Conference on Computer Vision and Pattern Recognition (CVPR), pp. 19989–19998 (2022)
68. Xiao, J., Shang, X., Yao, A., Chua, T.S.: Next-qa: next phase of question-answering to explaining temporal actions. In: Proceedings of the IEEE/CVF Conference on Computer Vision and Pattern Recognition, pp. 9777–9786 (2021)
69. Xu, D., Zhao, Z., Xiao, J., Wu, F., Zhang, H., He, X., Zhuang, Y.: Video question answering via gradually refined attention over appearance and motion. In: Proceedings of the 25th ACM International Conference on Multimedia, pp. 1645–1653 (2017)
70. Xuan, H., Zhang, Z., Chen, S., Yang, J., Yan, Y.: Cross-modal attention network for temporal inconsistent audio-visual event localization. In: Proceedings of the AAAI Conference on Artificial Intelligence, vol. 34, pp. 279–286 (2020)
71. Yang, A., Miech, A., Sivic, J., Laptev, I., Schmid, C.: Just ask: learning to answer questions from millions of narrated videos. In: Proceedings of the IEEE/CVF International Conference on Computer Vision, pp. 1686–1697 (2021)
72. Yang, P., Wang, X., Duan, X., Chen, H., Hou, R., Jin, C., Zhu, W.: Avqa: a dataset for audio-visual question answering on videos. In: Proceedings of the 30th ACM International Conference on Multimedia, pp. 3480–3491 (2022)
73. Yang, Z., Gan, Z., Wang, J., Hu, X., Lu, Y., Liu, Z., Wang, L.: An empirical study of gpt-3 for few-shot knowledge-based vqa. In: Proceedings of the AAAI Conference on Artificial Intelligence, vol. 36, pp. 3081–3089 (2022)
74. Yi, K., Gan, C., Li, Y., Kohli, P., Wu, J., Torralba, A., Tenenbaum, J.B.: Clevrer: collision events for video representation and reasoning. In: International Conference on Learning Representations (2020)
75. Yu, Z., Xu, D., Yu, J., Yu, T., Zhao, Z., Zhuang, Y., Tao, D.: Activitynet-qa: a dataset for understanding complex web videos via question answering. In: Proceedings of the AAAI Conference on Artificial Intelligence, vol. 33, pp. 9127–9134 (2019)
76. Yun, H., Yu, Y., Yang, W., Lee, K., Kim, G.: Pano-avqa: grounded audio-visual question answering on 360deg videos. In: Proceedings of the IEEE/CVF International Conference on Computer Vision, pp. 2031–2041 (2021)
77. Zadeh, A., Chan, M., Liang, P.P., Tong, E., Morency, L.P.: Social-iq: a question answering benchmark for artificial social intelligence. In: Proceedings of the IEEE/CVF Conference on Computer Vision and Pattern Recognition, pp. 8807–8817 (2019)
78. Zeng, K.H., Chen, T.H., Chuang, C.Y., Liao, Y.H., Niebles, J.C., Sun, M.: Leveraging video descriptions to learn video question answering. In: Proceedings of the AAAI Conference on Artificial Intelligence, vol. 31 (2017)
79. Zeng, A., Attarian, M., Choromanski, K.M., Wong, A., Welker, S., Tombari, F., Purohit, A., Ryoo, M.S., Sindhwani, V., Lee, J., et al.: Socratic models: composing zero-shot multimodal reasoning with language. In: The Eleventh International Conference on Learning Representations (2023)
80. Zhang, Z., Zhao, Z., Lin, Z., Song, J., Cai, D.: Localizing unseen activities in video via image query. arXiv preprint arXiv:1906.12165 (2019)
81. Zhang, H., Sun, A., Jing, W., Nan, G., Zhen, L., Zhou, J.T., Goh, R.S.M.: Video corpus moment retrieval with contrastive learning. In: Proceedings of the 44th International ACM SIGIR Conference on Research and Development in Information Retrieval, SIGIR '21 (2021)
82. Zhao, Z., Lin, J., Jiang, X., Cai, D., He, X., Zhuang, Y.: Video question answering via hierarchical dual-level attention network learning. In: Proceedings of the 25th ACM International Conference on Multimedia, pp. 1050–1058 (2017)
83. Zhao, Z., Yang, Q., Cai, D., He, X., Zhuang, Y., Zhao, Z., Yang, Q., Cai, D., He, X., Zhuang, Y.: Video question answering via hierarchical spatio-temporal attention networks. In: IJCAI, vol. 2, p. 8 (2017)
84. Zhao, W., Kim, S., Xu, N., Jin, H.: Video question answering on screencast tutorials. In: Proceedings of the Twenty-Ninth International Conference on International Joint Conferences on Artificial Intelligence, pp. 1061–1068 (2021)

85. Zhong, Y., Ji, W., Xiao, J., Li, Y., Deng, W., Chua, T.S.: Video question answering: datasets, algorithms and challenges. In: Proceedings of the 2022 Conference on Empirical Methods in Natural Language Processing, pp. 6439–6455 (2022)
86. Zhou, L., Xu, C., Corso, J.: Towards automatic learning of procedures from web instructional videos. In: Proceedings of the AAAI Conference on Artificial Intelligence, vol. 32 (2018)
87. Zhou, Y., Guo, A., Huo, S., Liu, Y., Kung, S.Y.: Weakly supervised video re-localization through multi-agent-reinforced switchable network. IEEE Transactions on Circuits and Systems for Video Technology (2023)

Chapter 4
Future Research Directions

Abstract The classical *Video Grounding* setting has received abundant study, and state-of-the-art models are now capable of achieving remarkable performance. Nevertheless, there remain several significant issues to be explored. In this chapter, we aim to introduce these directions, including out-of-distribution *Video Grounding*, explainable *Video Grounding*, and the application of vision-language pre-training to *Video Grounding*. Preliminary investigations have been conducted on some of these topics, so we will summarize the early works and point out possible future directions simultaneously.

4.1 Out-of-Distribution Video Grounding

Out-of-distribution (O.O.D.), or bias, is a topic that is emerging in various deep learning areas [25]. With the development of Video Grounding methods, it has been found that the components of VG datasets are also biased, raising questions on out-of-distribution Video Grounding. This task concentrates on addressing several patterns that cause biases in Video Grounding, which include annotation distribution shift, query distribution shift, visual activity distribution shift, and joint distribution shift.

4.1.1 Out-of-Distribution as a Generically Emerging Topic

In this section, we aim to first establish some background knowledge of O.O.D. so that the following sections of O.O.D. in *Video Grounding* are well-contextualized among all related studies.

Currently, most deep learning methods rely heavily on sufficient high-quality and sufficient training data. Furthermore, most methods are based on the assumption that the training and testing data are from the same data distribution, i.e., I.I.D. assumption. However, this is not always the case in real applications.

X. Wang et al., *Video Grounding and Its Generalization*,
https://doi.org/10.1007/978-3-031-94837-4_4

To be specific, it is usually inevitable that there are *biases* in a dataset. Taking image classification as an example, the number of images for each class may not be the same (also known as long tail [51]), and the images for one class may contain some unintended similarity. For instance, all images of dogs have lawns in the background. This property can also be described as *imbalanced distribution* of data. Such biases can cause a model to learn unintentionally incorrect knowledge from the given dataset, such as incorrectly assuming that all dogs are on lawns. Biases can also take other forms, such as image style, geographic location, or camera parameters [25], thus posing a widespread issue in deep learning.

These biases, though potentially harmful, can not be easily exposed if the training and testing sets share the same problems. Hence, it is now a common practice to intentionally design such a dataset in which the data distribution in the testing split differs from the training distribution [51]. This setting can also better reflect real-world scenarios, thus proving great importance if we aim to design a model truly applicable.

Given the basic setting of O.O.D., how can it be tackled?

1. Without introducing new data, it is viable to adjust the training data and reduce bias.
2. Another approach is to design special model architectures or training policies to force the model to learn correctly from a biased dataset.
3. The emerging large base models are considered to contain strong and unbiased knowledge, and integrating them is also a solution to the O.O.D. problem.

Next, we will provide a more detailed explanation of these methods, and Fig. 4.1 illustrates both the cause of the O.O.D. problem and its three solutions.

Dataset Debiasing Since data is the root of O.O.D., directly adjusting the dataset is an intuitive solution. However, introducing new data to reduce the data bias can be too expensive, so we aim to debias a dataset based solely on the original data without introducing any additional data. Hence, these methods share some similarities with data augmentation but with different purposes. A good example is imbalanced classification [21], where the number of samples to each class is imbalanced. A natural solution is to re-sample the samples till the numbers for each class become identical, which includes somehow generating new samples for minority classes and discarding random majority samples. Since the intuitive approach of simply repeating minority samples is not yielding satisfying results, further improvements are targeted for better strategies for sample generation, such as generating more representative and harder samples [49]. Another equivalent approach is class re-weighting while calculating losses, which basically assigns larger weights to minority classes and smaller weights to majority ones, such as class-balanced loss [11] and Focal Loss [48]. However, dataset debiasing methods also possess drawbacks: a dataset debiasing method needs to explicitly define what is considered a bias in a dataset, and thus, it is possible for it to overlook some biases. In fact, the difference between the numbers of samples is only a very simple case, while there are more biases that are harder to notice and define in real applications.

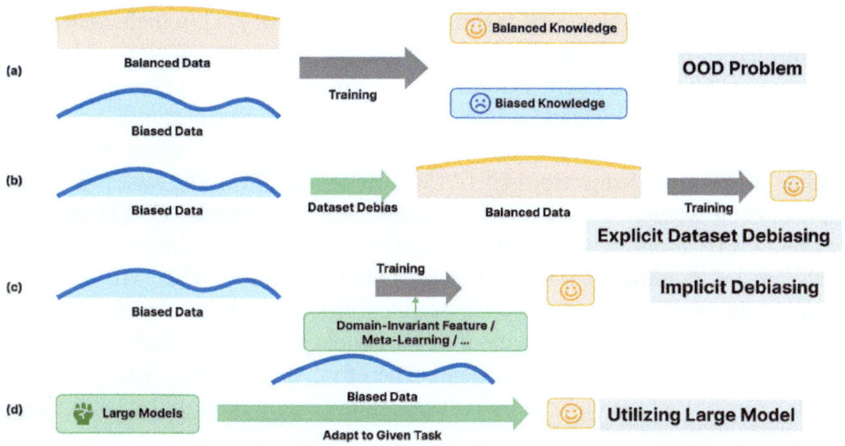

Fig. 4.1 Visual illustration of O.O.D. concept and typical methods. (**a**) O.O.D. problem originates from biased training data. (**b**) Explicit dataset debiasing method: directly modifying training data to reduce biases. (**c**) Implicit debiasing: designing special model architecture or training policies so that the model can learn balanced knowledge from biased training data. (**d**) Utilizing large model: adapting pre-trained large base models, which are considered to possess strong and balanced knowledge, to a given task with a small amount of biased data

Implicit Debiasing via Model Architecture or Training Policy Having known the drawbacks of dataset debiasing methods, other methods are being proposed. If explicit reduction of manually-defined biases can cause degradation, why not implicitly reduce biases? This type of approach includes special model architectures or training policies to force the model to learn implicitly debiased knowledge. Some methods of this type change the aforementioned re-sampling and re-weighting methods into an implicit fashion. For example, Li et al. [44] model the distribution shifts with uncertainty and utilize dataset statistics to alleviate the domain perturbations, while Zhang et al. [91] propose to learn the weights of training samples. There are also methods that are less related to explicit dataset debiasing, such as Risk Extrapolation [38], an optimization method to minimize the differences in risk across multiple training domains. Another way is to learn domain-invariant features by adversarially preventing the learned features from being discriminative for domain shifts [17]. Implicit debiasing methods also have certain drawbacks: they usually cause additional complexity that may affect training efficiency, making it hard to judge if a method is truly effective.

Integration of Large Base Models O.O.D. originates from the inner bias and insufficiency of a single dataset, so introducing more data into the training set is likely to be a good solution. Although gathering more data is often hard, the emergence of large base models has provided a workaround. More specifically, large models have been trained on significantly vast data. And by adapting a pre-trained large model to a downstream task, we can utilize its training data for the given task. This is the prevalent approach in the age of large base models, and there are abundant examples. A very famous model for this purpose is CLIP [71], which contrastively learns the cross-model similarity between images and texts on web-scale image-caption pairs. CLIP, by itself, can already perform zero-shot tasks in real-world scenarios, such as image classification in a novel domain. Furthermore, when a task is too complex to utilize CLIP directly, it is a common practice to fine-tune the model or distill knowledge from CLIP [56]. Benefiting from the vast pre-training, CLIP has been widely adopted in many tasks, including various O.O.D. scenarios. However, CLIP is already a relatively old model, whereas abundant new large models present significant possibilities for O.O.D. learning. This will be covered in detail in the last section of this chapter. Integrating large base models is based on the assumption that the training data of large models is well-balanced and sufficient, but this assumption doesn't always stand. Therefore, the first two types of methods can still be useful, given the impressive performance gain brought by large models.

Besides the developments introduced above, there have also been attempts to understand the O.O.D. problem on a theoretical basis. Ye et al. [84] have provided a framework that formally defines O.O.D. and its learnability, introducing a new concept of expansion function, and proving O.O.D. generalization bounds. This framework naturally induces a model selection criterion to complete the algorithm. Such efforts can build a solid ground for future O.O.D. studies.

More specifically, for the *Video Grounding* task, a data sample consists of a ground truth moment annotation, a query, and a video. Thus, each of the three components can face the distribution shift independently, and we can also consider the shift of their joint distribution.

Video Grounding in O.O.D. setting is still an ongoing research topic. In the following parts, we will provide the basic definition of each distribution shift setting and related work.

4.1.2 Annotation Distribution Shift

Definition and Datasets Annotation distribution shift is the first to be investigated among all O.O.D. settings [65, 89]. It is found that the ground truth moment annotations in *Video Grounding* datasets are biased, concentrating on a certain pattern. More importantly, these biases are identical in training and testing data, resulting in a model's high prediction when it actually lacks cross-model alignment.

More specifically, Fig. 4.2 shows the moment annotation distribution of Charades-STA and ActivityNet Captions. It is clear that they are highly concentrated on just a few small regions. Furthermore, in [89], a bias-based model is designed to simply generate predictions based on the statistical bias without access to either query or video. This simple method, however, outperforms several state-of-the-art models, demonstrating the necessity of dealing with the annotation distribution shift.

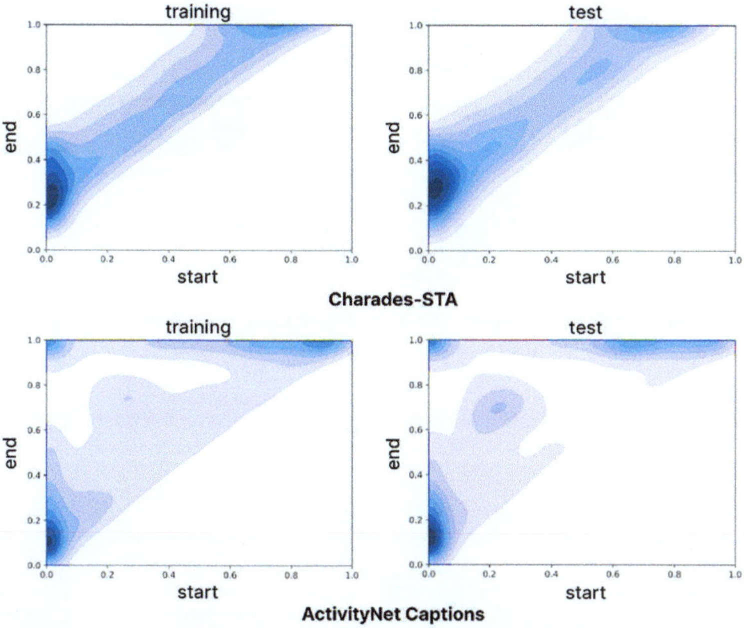

Fig. 4.2 Visualization of moment annotation distribution of Charades-STA and ActivityNet Captions. Deeper color indicates higher density

Yuan et al. [89] provide re-splits of Charades-STA and ActivityNet Captions, namely test-iid and test-ood, for the research on annotation distribution shift. Later, a new dataset, MAD [74], is designed to alleviate distribution bias. Currently, re-splits of Charades-STA and ActivityNet Captions are still the most popular datasets, while MAD has not been widely adopted.

Researchers' attempts to address the annotation distribution shift can be categorized into **model-based approaches** and **training-strategy-based approaches**. They are orthogonal to each other and can be combined to reach better results. Moreover, it is possible to edit training data to get unbiased samples, as shown in Fig. 4.3. However, such samples are less realistic and may have temporal consistency damage, so it is not advisable to simply use them as augmented data to train a model. Therefore, this approach might be adopted as part of other methods but not be used independently.

Model-Based Methods To begin with, the core concept of model-based methods is to use unimodal branches to deliberately capture biases and then manage to disentangle them from the full model. Zhang et al. [90] adjust losses dynamically, as shown in Fig. 4.4. More concretely, element-wise multiplication is conducted between the loss of the full model and the losses of unimodal branches to adjust element weights so that biases are prevented from being learned by the full model. A similar approach is adopted in [5], where the prediction of the unimodal branch

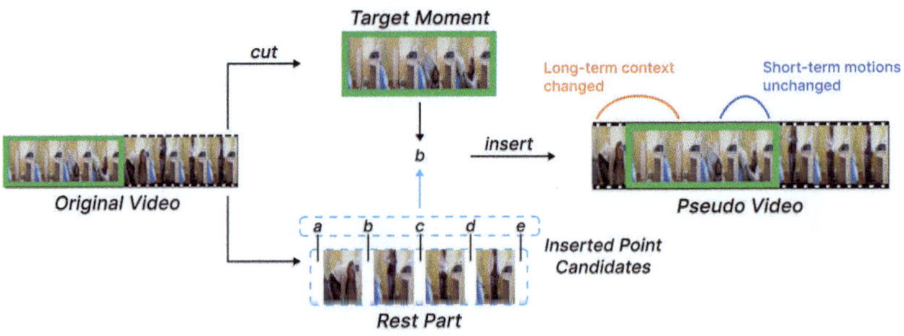

Fig. 4.3 One way to edit data samples. The ground truth moment is taken away from its original position and inserted back into the rest of the video at a random position

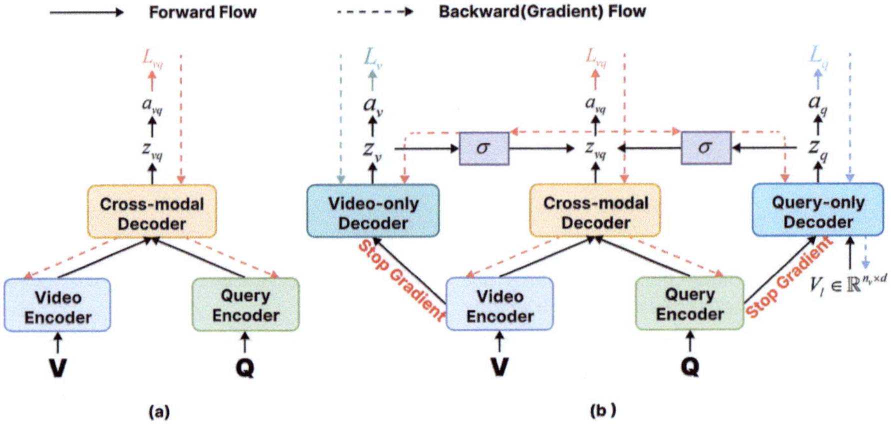

Fig. 4.4 The framework of [90]. When a sample is answered with high accuracy by unimodal branches, the element-wise multiplication will prevent the full model from learning the distribution bias. (**a**) Standard Proposal-free TSGV Model. (**b**) Proposal-free TSGV Model with Model Debiasing (MD)

is used to reweigh the importance of the training samples, suppressing samples containing biases, as shown in Fig. 4.5. Liu et al. [53] adopt a different approach, which focuses more on the intermediate representations rather than losses, as shown in Fig. 4.6. In their approach, the intermediate representations of unimodal branches are subtracted from the representation of the same layer from the full model, forcing the model to learn debiased knowledge.

Other model-based methods take the perspective of causality and achieve debiasing with causal intervention. Yang et al. [82] propose the first causality-based *Video Grounding* method, as shown by Fig. 4.7. This method builds a structural causal model to capture the causality between different *Video Grounding* components while avoiding harmful confounding effects caused by moment location. Lan et al. [39] designs a Multi-branch Deconfounding Debiasing (MDD) framework that

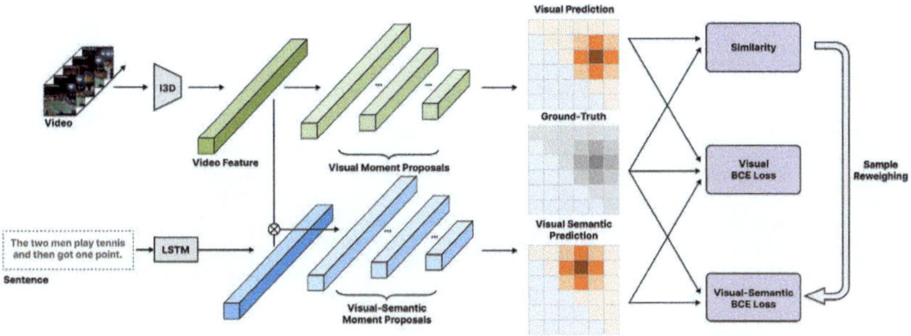

Fig. 4.5 The framework of [5], where the similarity between predictions of the two branches is used to reweigh samples

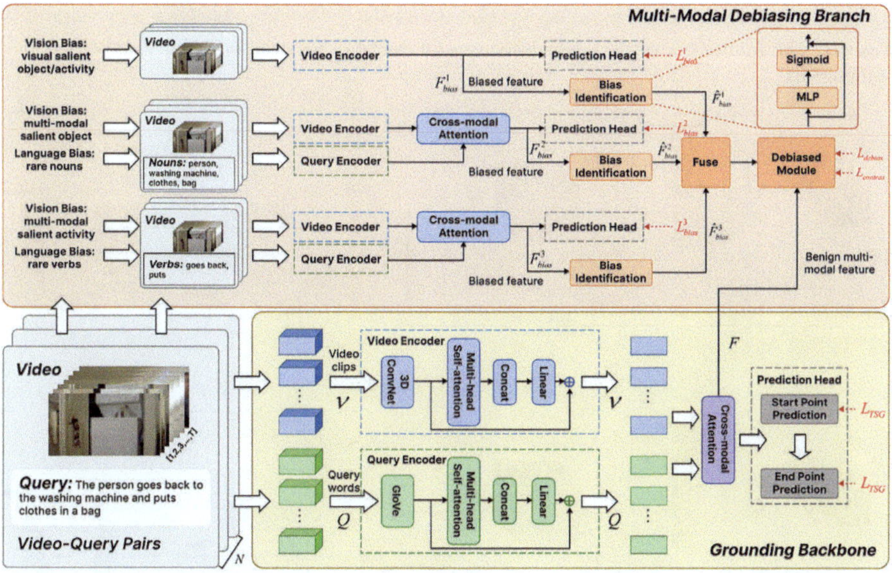

Fig. 4.6 The framework of [53], where several branches are designed to capture data bias, and their representations are subtracted from the corresponding representation in the full model

contains semantic role tree, position reconstruction, and multi-branch deconfounder, as shown in Fig. 4.8.

Training-Strategy-Based Methods To address the distribution shift of ground truth annotations in *Video Grounding* datasets, another solution is to carefully design the training strategies for the *Video Grounding* models to adapt the ground truth bias. Hao et al. [23] use aforementioned unbiased data via editing in two auxiliary tasks, as shown in Fig. 4.9. One is cross-model matching, requiring the model to predict frame-level cross-modal relevance and ensure its consistency even when temporal positions change. The other is temporal order discrimination, requiring the model

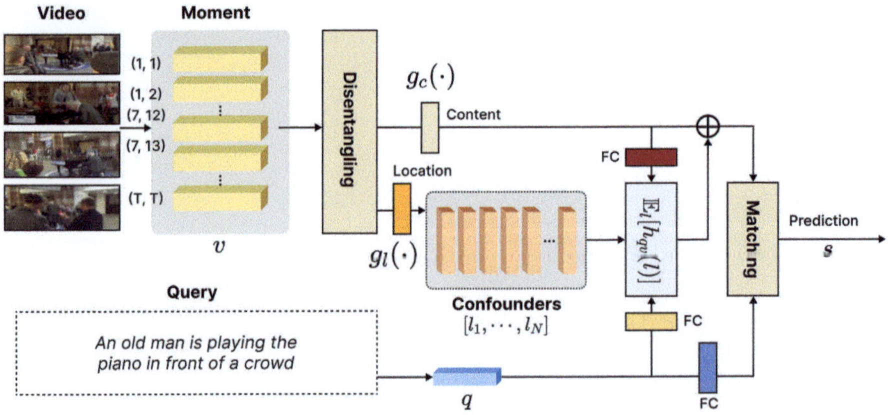

Fig. 4.7 The framework of [82], which focuses on causality between different *Video Grounding* components

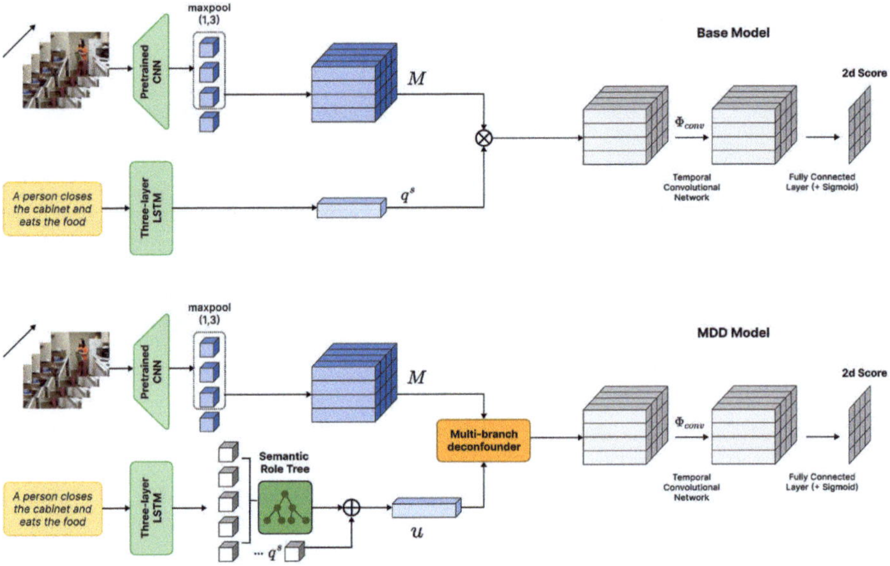

Fig. 4.8 The framework of MDD, adding semantic role tree, position reconstruction, and multi-branch deconfounder to a base model for the sake of causality

to discriminate if the moment sequence is in the correct order. They strengthen the model's perception ability on both short-term and long-term information.

Lan et al. [40] use multiple negative augmentations in a hierarchical way in order to generate unbiased data, and further utilize generated data in a curriculum learning framework. More specifically, as shown in Fig. 4.10, they train the model in a contrastive manner where generated pseudo data is used as negative samples.

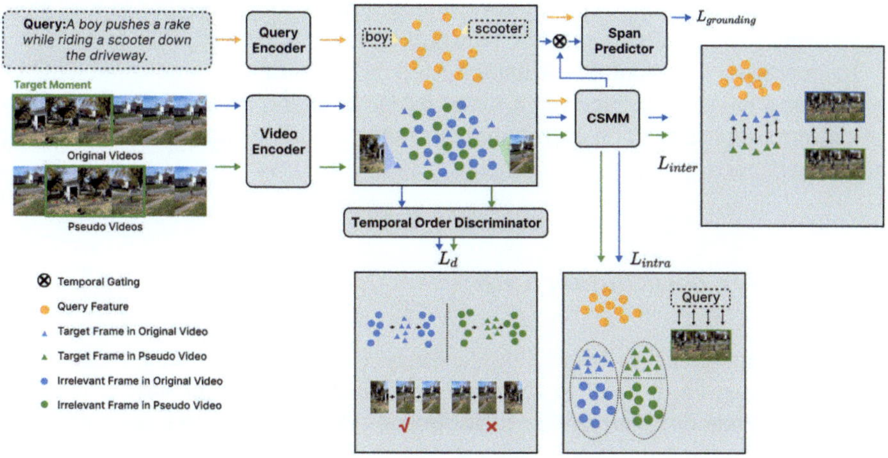

Fig. 4.9 The framework of [23], where two auxiliary tasks are added to the training procedure to improve the model

Fig. 4.10 The framework of [40], where the left side demonstrates the contrastive learning procedure, while the right side shows the multi-stage curriculum learning process

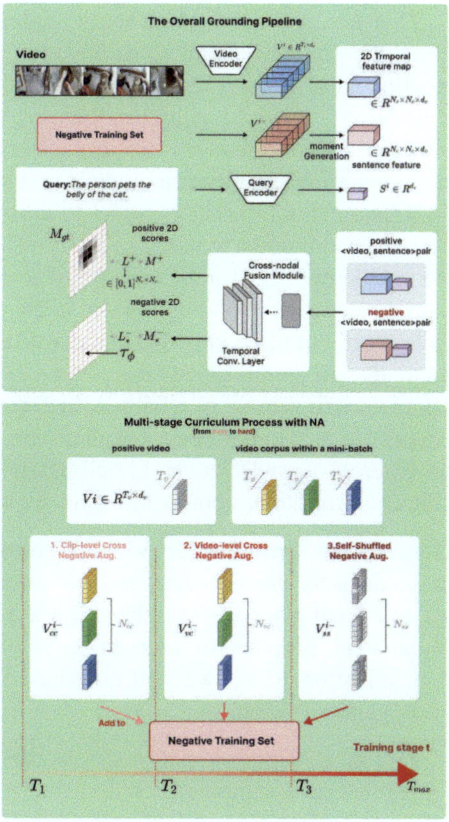

Furthermore, a multi-stage curriculum learning framework is designed to start from easy samples and gradually move to hard samples.

To conclude, annotation distribution shift is the simplest among all O.O.D. settings and has already received a lot of research. In the future, more focus may be shifted from annotation distribution to other O.O.D. settings, which will be introduced later.

4.1.3 Query Distribution Shift

Definition and General Idea Since a query is a complex sentence in natural language, it can be difficult to accurately define its distribution shift. Therefore, we tend to focus on a slice of it: novel events (unseen concepts) generalization, which is also a current research interest. Events or concepts are typically objects and activities in queries. For example, the queries "the person pours some water into the glass" includes "person", "pour", "water" and "glass". Based on this, novel event generalization requires that the model performs well on testing samples which have events not included in training data.

This setting is, in fact, strongly related to the real-world applicability of *Video Grounding* models. Training data, which requires the heavy labor of collecting and labeling data, is always limited in size and diversity. As a result, if we want to apply a model to real-world scenarios, there are inevitably novel events. With this demand in mind, we can manually create a testing set with novel events to better simulate such scenarios and reflect a model's real-world applicability.

As we cannot expect a model to correctly process a sample containing events previously unknown to it, the key to novel event generalization is to provide suitable external knowledge for the model as reference. Two typical ways of providing external knowledge are pre-training and knowledge graphs, as representative approaches to capturing implicit knowledge and explicit knowledge respectively. More detailed methods will be discussed later.

An easier setting is what's called compositional generalization, where every concept is previously seen, but their composition is novel. This setting relies less on external knowledge, as a well-designed model should be able to disentangle concepts from their composition and acquire enough knowledge through training.

Generalized Datasets for Video Grounding Previous *Video Grounding* benchmark datasets are not designed for novel event generalization, as their training set and testing set are sampled from the same data distribution. Therefore, some researchers propose to reorganize splits of previous datasets for novel event generalization. Here, we will introduce some datasets that can be used for novel event generalization.

Paul et al. [66] use spaCy [27] to parse verbs and nouns from text queries and reorganize data samples according to them in order to let the testing set contain

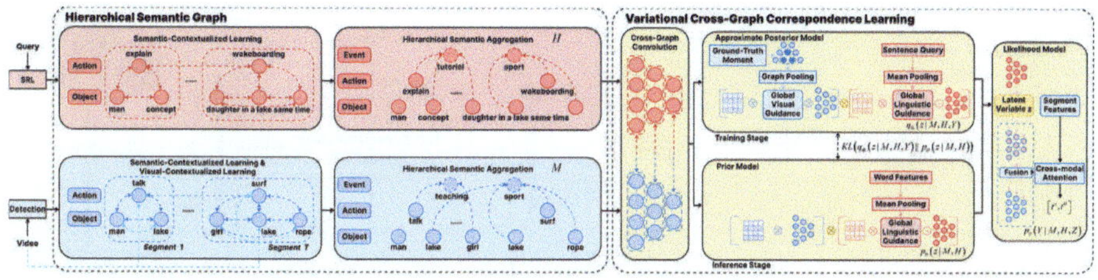

Fig. 4.11 Overview of VISA framework, consisting of a hierarchical semantic graph and a variational cross-graph correspondence learning component

unseen events. As a result, they propose two new datasets, namely Charades-STA Unseen and ActivityNet Captions Unseen.

Li et al. [45] provide another dataset, called Charades-CG and ActivityNet-CG, where CG stands for compositional generalization. This dataset contains three testing splits: test-trivial for I.I.D. testing, novel-composition for compositional generalization testing, and novel-word for novel event generalization testing.

Compositional Generalization Methods To address the query distribution shift, some researchers attempt to disentangle concepts from the composition of Video Grounding data and acquire knowledge through training with their well-designed models. Li et al. [45] design VISA for compositional generalization problem, as shown in Fig. 4.11. Its key philosophy is to disentangle individual concepts from their composition through a hierarchical semantic graph. VISA first parses the given video and query and decomposes them into three hierarchies, i.e., global events, local actions, and atomic objects. VISA then conducts cross-modal interaction upon the semantic graph instead of the original video and query. To infer the fine-grained cross-graph correspondence, VISA models it as a latent variable z and formulates the problem into a variational inference framework. This method is further extended in [47], where an Adaptive Structured Semantics Learning approach is added, encouraging the model to capture the structure-informed and domain-generalizable semantics (Fig. 4.12).

Yang et al. [83] argue that composition representation should be sensitive to different levels of granularity, and propose a new method for compositional generalization. Different levels of granularity are achieved by decomposing queries into several sub-sentences. Since there is no corresponding ground truth for these sub-sentences, this model is trained in a weakly supervised manner via decomposing and reconstruction.

Paul et al. [66] propose TLRR for novel event generalization, as shown in Fig. 4.13. Their basic hypothesis is that there exist some similarities between the corresponding moments of semantically relevant queries. Therefore, when dealing with an unseen query, it is possible to first find a relevant observed query and use it as the support moment to predict the target moment. To implement this procedure, TLRR utilizes a universal sentence encoder to find a relevant query, and it samples

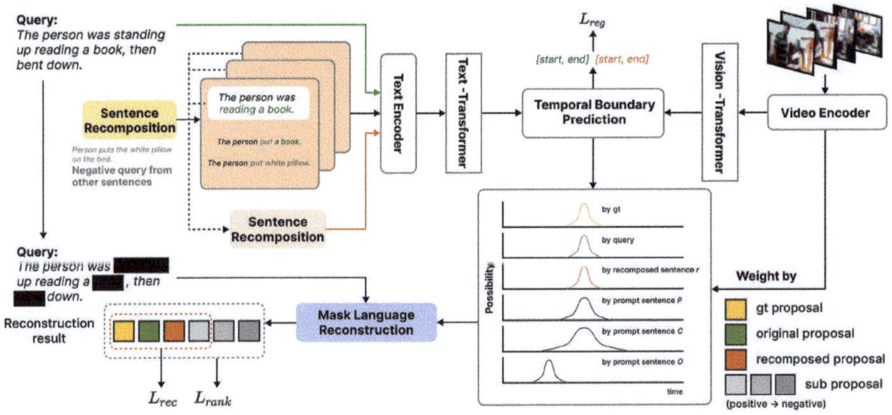

Fig. 4.12 Overview of DeCo framework. A query sentence is decomposed into sub-sentences, and they are used to construct compositional sub-sentences and negative subsentences

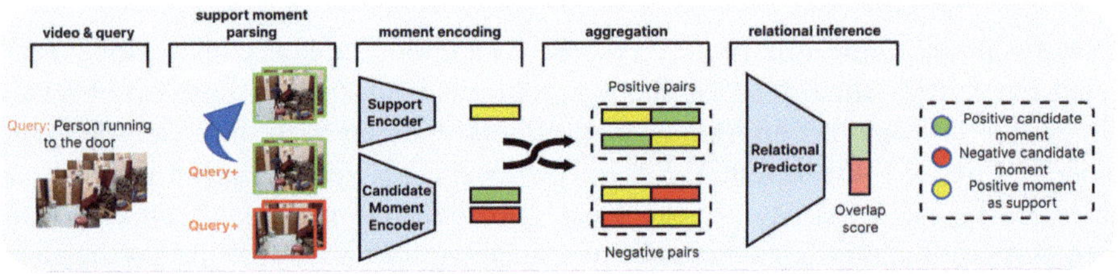

Fig. 4.13 Overview of TLRR framework. To train the relational reasoning module, candidate moment and support moment are aggregated to form positive pairs and negative pairs, based on which the module is trained to estimate the relational scores

several candidate moments, modeling the calculation of overlap score as relational reasoning.

Luo et al. [57] propose a vision-language pre-training framework, namely VDI, to improve *Video Grounding* models' ability to generalize unseen events, which is shown in Fig. 4.14. They observe that conventional *Video Grounding* approaches usually adopt pre-trained unimodal encoders, but unimodal encoders are not sufficient to make the *Video Grounding* model generalizable. Therefore, vision-language pre-training is more likely to empower *Video Grounding* models, but neither models trained on coarse-grained video-level labels nor image-based pre-training models are satisfiable. Hence, VDI injects the visual context and spatial dynamic information from videos into text embeddings.

Jung et al. [35] design BM-DETR to overcome the weak visual-textual alignment problem, as shown in Fig. 4.15. They state that the weak visual-textual alignment restricts further performance gain of previous *Video Grounding* methods, and utilizing negative queries can lead to better overall alignment in videos. More concretely, this model calculates the joint probability of a target query and the

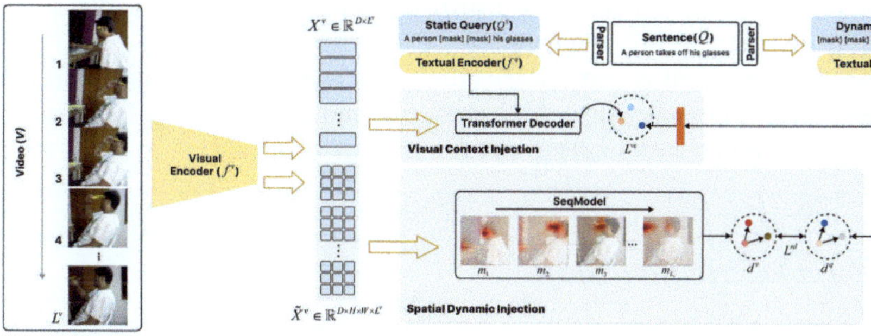

Fig. 4.14 Overview of VDI framework, where the Visual Context Injection and Spatial Dynamic Injection modules capture the visual context and spatial dynamic information, respectively

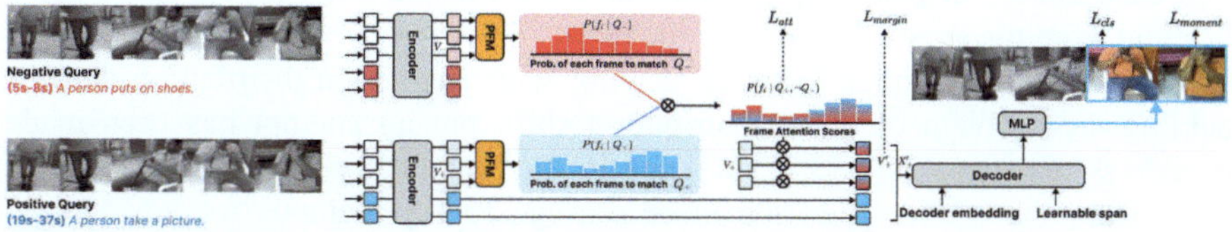

Fig. 4.15 Overview of BM-DETR framework. The joint probability of both target query and negative query for each frame is calculated to provide frame attention scores, which are used to refine multimodal features

complement of negative queries for each frame and uses the result to update multimodal features and predict the target moment.

In summary, query distribution shift is the current research hot spot of the *Video Grounding* community. There is already much work on this topic, yet these methods are pretty diverse, indicating that we have not reached a consensus on a best approach. Meanwhile, we notice that the performance of state-of-the-art methods is still far from being optimal. As a result, this direction may continue to be the research focus for some time.

4.1.4 Visual Activity Distribution Shift

Theoretically, we should also consider the visual activity distribution shift. However, if a query contains novel events, there must be corresponding novel events in the ground truth moment as well. Therefore, the distributions of queries and visual activities are not entirely independent.

Nevertheless, we can still imagine a situation where novel events exist only in moments that are not ground truth, in order to separate visual activity distribution shift from query distribution shift. In this case, the model is required to know that a moment is **not** the target moment even if it contains novel events. Intuitively, this should be an easier problem than query distribution shift, yet there has not been any focus on it to demonstrate this judgment.

In summary, the visual activity distribution shift is not of great interest to the *Video Grounding* community. It is safe to skip it and move directly to the joint distribution shift, which we will discuss next.

4.1.5 Joint Distribution Shift

Lastly, we should consider the joint distribution shift of all three components of a *Video Grounding* data sample as the ultimate form of O.O.D. setting.

In this setting, distribution shift occurs in both annotation and query, interacting with each other. For instance, the annotation shift of the event "wake up" may be different from the annotation shift of the event "drink water".

Current methods either focus on dealing with annotation distribution shift or tackling the problem of query distribution shift, but no attempt has been made to address their co-occurrence. As a result, the method for one setting may still suffer from the challenge of the other setting. Taking methods for novel event generalization as an example, they usually rely on external knowledge to generalize to novel events, but this does not necessarily grant them the required ability for annotation O.O.D. setting.

In order to address the challenge of joint distribution shift, it might be beneficial to refer to the research of domain generalization (DG). As shown in Fig. 4.16, each domain is composed of data sampled from a certain distribution. More formally, let P^i_{XY} be domains, then (x^i_j, y^i_j) is data consisting all domains. Typically, multiple domains are accessible during training, and the target is to train a model that can generalize to the unseen testing domain. DG methods include data augmentation, domain invariant representation, and learning strategies [79].

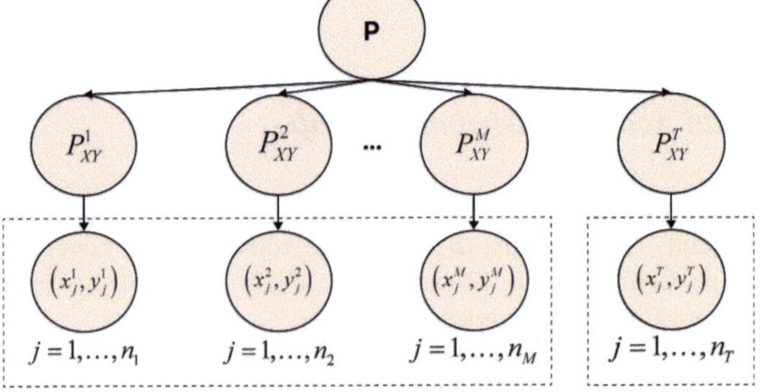

Fig. 4.16 Illustration of domain generalization, which aims to train a model that can generalize to the unseen testing domain based on multiple training domains

Data augmentation aims to generate new training samples by combining some training domains together [79]. Hopefully, these new samples can partially cover the distribution of the testing domain, thus improving the performance.This seems to be a promising approach. However, the joint distribution shift setting requires generating videos containing novel events as well as high-quality natural language queries, which is non-trivial.

Domain invariant representation aims to find the same representation for a concept [79]. A similar method is used in a *Video Grounding* work [53]: they explicitly capture the biased representation and remove it from the original representation.

Learning strategies are special training methods that improve models' generalizing ability, including ensemble learning, meta-learning, gradient operation [79] etc.

Although DG methods generally look helpful, adapting them to the *Video Grounding* task can be difficult, as *Video Grounding* is much more complex than the commonly used DG tasks such as image classification.

4.2 Explainable Video Grounding

Early-stage Video Grounding simply focuses on applying deep learning methods for training to more accurately localize video segments based on the grounding query. However, these methods are usually black-box and lack the interpretability to reveal deep-level mechanisms for why the models localize grounding segments correctly or incorrectly. Therefore, several research works attempt to make the Video Grounding models more **explainable**, providing an intuitive understanding of model mechanisms and predictions and further improving the model performance, which includes applying causality-based Video Grounding, modular neural networks, and neural symbolic reasoning.

4.2.1 Causality-Based Video Grounding

Causal inference aims to capture the true causality in a task, preventing models from simply fitting the confounding pattern and dataset bias. It is thus helpful for explainable video grounding and has already been adopted in some *Video Grounding* methods. Most of them are motivated by annotation distribution shift problems and are already discussed in Sect. 4.1.2. We will provide a general introduction to causality, followed by the rest of causality-based *Video Grounding* methods.

For explainable algorithms, association reveals the relation between objects, leading to better explainability. While the association is an important progress, causality takes a step forward and tries to understand the true causes and effects [81]. Therefore, causality-based methods can exploit the process for data generation instead of surface predictive or descriptive patterns. A more detailed concept for

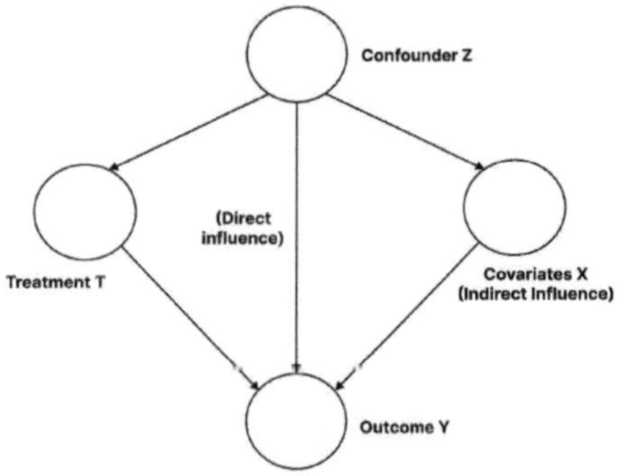

Fig. 4.17 Conceptual illustration of four concepts in causality analysis. Confounder Z influences both Treatment T and Outcome Y, making the causal influence more difficult to evaluate

causality is causal inference, which estimates the causal effects of inputs on the outcome. As such, causal inference can benefit video grounding, providing a more reasonable and explainable inference process.

Formally, a causal model consists of Covariates X, Confounder Z, Treatment T, and Outcome Y, where we aim to evaluate the true causal inference of T on Y. The tricky part of the model, as shown in Fig. 4.17, is that Confounder Z has an influence on both Treatment T and Outcome Y (both directly and indirectly through X), making it difficult to evaluate the true causal influence. We can use structural equations to represent the aforementioned relation:

$$\begin{aligned} Z &= f_Z(E_Z) \\ T &= f_T(Z, E_Y) \\ Y &= f_Y(Z, T, E_Y), \end{aligned} \tag{4.1}$$

where E_Z, E_Y, E_Y are external conditions independent of variables in the causal model.

Suppose a binary treatment scenario where T is either 0 or 1. We can observe the corresponding outcomes given each assignment to T, as Y_0 and Y_1. The individual treatment effect (ITE) is defined as how outcome changes with covariates X fixed. Suppose X is fixed as X_i, the corresponding ITE is:

$$ITE(X_i) = E[Y_1|X_i] - E[Y_0|X_i]. \tag{4.2}$$

If we observe a group of samples with different X_i and take the average, the formulation becomes average treatment effect (ATE):

$$ATE = E[Y_1 - Y_0]. \tag{4.3}$$

To analyze the relationships between the parameters of a neural network, intervention is a feasible method. Suppose we want to evaluate the causal attribution

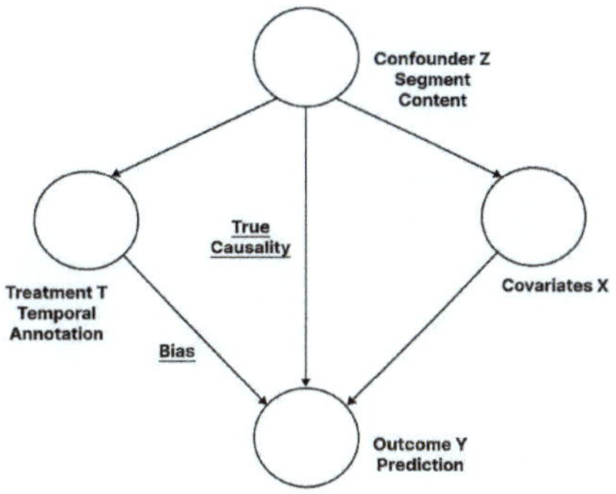

Fig. 4.18 An example for modeling video grounding as causal inference, where bias comes from the duo influence of confounder (video segment content) on temporal annotation and model prediction

of neuron x_i to output neuron y_i, we can manually fix the value of x_i as $do(x_i = \alpha)$ and observe how y_i changes:

$$\mathbb{E}[y|do(x_i = \alpha)] = \int yp(y|do(x_i = \alpha))dy. \tag{4.4}$$

If the relationships of parameters are complicated and intervention thus becomes infeasible, we can alternatively use post-hoc interpretation methods. For example, Casual Feature Learning instead evaluates the causal relation between learned features and model predictions, and Counterfactual Explanation generates new data samples that can change the model's predictions.

The introduction so far has shown model-agnostic interpretation methods, which can be applied to any neural networks that have already been trained to interpret their inner mechanics. Naturally, for a given *Video Grounding* model, it is possible to apply such model-agnostic interpretation methods and evaluate if the model has truly learned the multimodal knowledge. Besides directly applying model-agnostic methods, it is also possible to devise video grounding models inspired by causality. To this end, it is important to set up the causality model correctly according to the settings of video grounding, which might be non-trivial. An example of modeling video grounding as causal inference is shown in Fig. 4.18. We may regard the prediction's correctness as the outcome, the actual content of a video segment as the confounder, and the distribution of the corresponding temporal annotation as the treatment or a covariate. In this example, a non-causal method can incorrectly fit the distribution of temporal annotations, while a causal method has the potential to correctly understand the actual content of videos.

Next, we provide additional examples of causality-based *Video Grounding* methods. Nan et al. [64] propose a model with both intervention and dual contrastive learning to overcome dataset bias. This method, as shown in Fig. 4.19, first uses two contrastive modules to improve representation quality, and then uses causal interventions with events as surrogate confounders to learn disentangled representations.

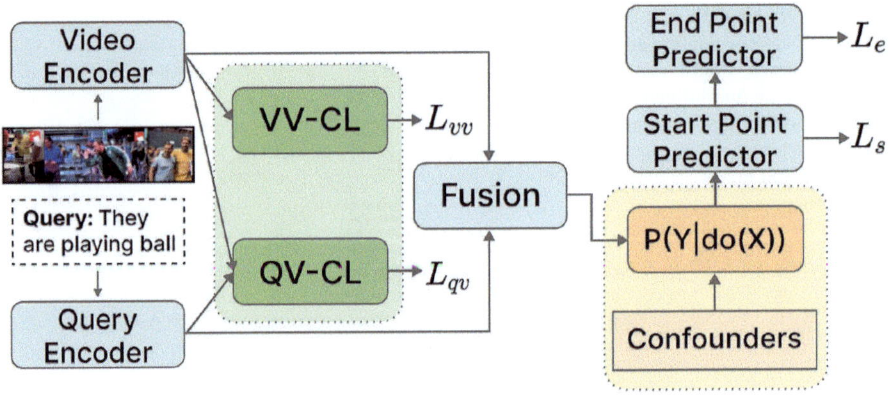

Fig. 4.19 The framework of IVD-DCL, which has two contrastive modules (i.e., VV-CL and QV-CL) together with a causal confounder, compared with base *Video Grounding* models

4.2.2 Video Grounding via Neural Module Networks

Neural Module Network (NMN) [15] is a good option for tasks that are compositional by nature. For example, a visual question-answering task can be decomposed into object detection, object classification, relation inference, and so on. For a compositional task like the example, NMN decomposes it into elemental units that can no longer be decomposed and then implements each elemental unit as a module. Typically, such a composition is dynamic, as NMN needs to first parse the given task of a data sample and select modules accordingly. With its compositional nature, NMN excels at explainability in two ways. On one hand, though each module is not naturally explainable, it is much simpler than complicated neural networks, thus being able to be analyzed easily with post-hoc interpretation techniques such as saliency maps. On the other hand, the high-level composition of elemental modules is explainable, so NMN as a whole can offer satisfying explainability. If we aim for better explainability at tasks which are more complex than simpler tasks like classification, such as question answering and expression referrals, NMN serves as a good candidate for both performance and explainability.

Conventional NMN methods preceding the deep learning era [4] typically employ pre-defined and static modules, including many sub-task modules and a decision-making module to summarize the sub-results and make the final decision. The overall composition of modules remains the same for all data samples, which makes these methods different from newer approaches. In contrast to conventional approaches, modern neural module networks tend to re-structure the whole network adaptively for every given instance, taking a step forward. Specifically, three major components are utilized: module inventory, module selector, and module assembler, as shown in Fig. 4.20. The module inventory contains a few pre-defined modules, where each module is a simple neural network, capable of its assigned task. The module selector usually selects modules based on linguistic clues, such as the

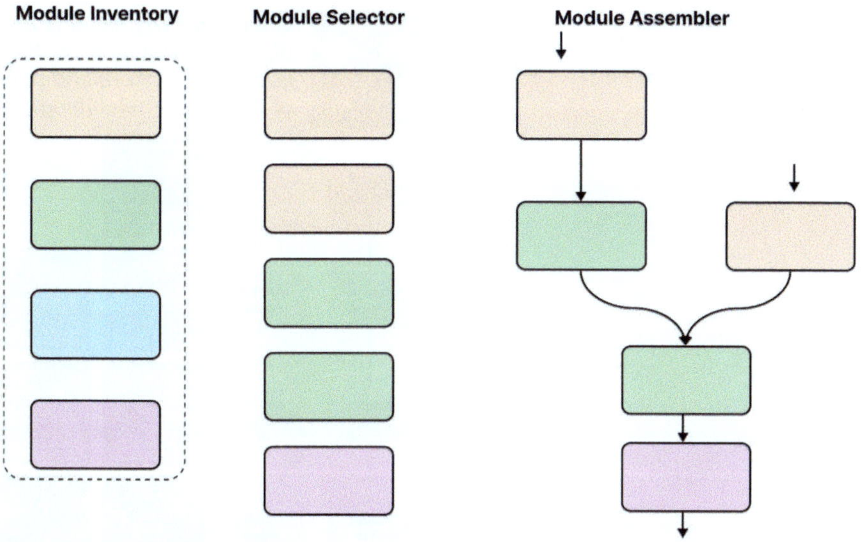

Fig. 4.20 Modern neural module networks are typically implemented as a module inventory, a module selector, and a module assembler

question in natural language, for the case of visual question answering. This module selector can either be a rule-based design or a neural network to be more flexible. Finally, the module assembler composes the selected modules into a complete end-to-end network, and it should be noted that the order of modules matters as the modules depend on the contexts. Furthermore, the sampling of modules can either be discrete or soft, where soft sampling means each module is assigned a weight, indicating its contribution to the output.

As stated before, NMN suits tasks that are complex and decomposable in nature, which is the nature of *Video Grounding* tasks. Therefore, if we want to grant *Video Grounding* models better explainability, NMN is a potential choice. However, there have not been many attempts at it, so we will first focus on NMN-based VQA as a methodology reference.

The task of VQA can be decomposed into object detection, object classification, attribute classification, counting, and relationship detection [15]. This indicates the rationale of NMN-based VQA approaches. Mascharka et al. [62] propose the Transparency by Design (TbD) model to address the gap between performance and interpretability. Their model consists of modules including Attention, Query, Relate, Same, Comparison, And, and Or, which are dynamically composed given a certain question. An example is displayed in Fig. 4.21. The key innovation comes with how information is exchanged between modules: they force modules to output and take in visual attention masks instead of high-dimensional feature maps in previous methods.Meta Module Network (MMN) [7] is another modular method for VQA, as shown in Fig. 4.22. MMN translates questions into programs, instantiating programs

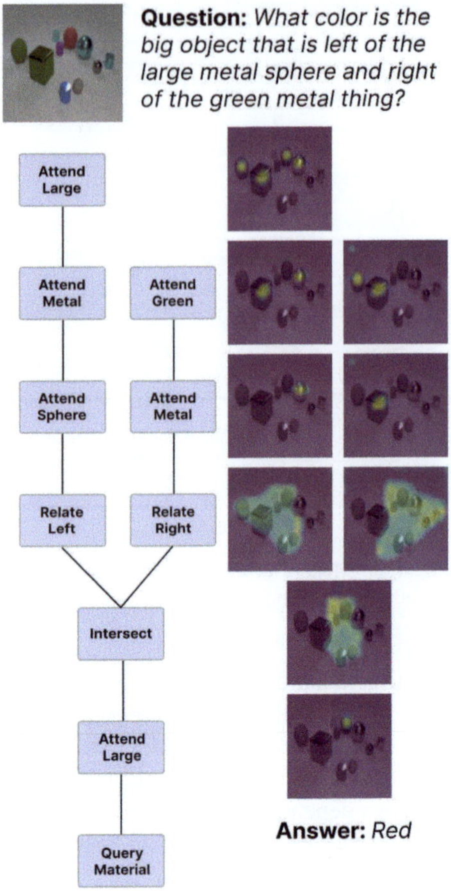

Fig. 4.21 The TbD model is dynamically composed of pre-defined modules based on the given question

as modules, and executing modules to get the predictions. For other NMN-based VQA approaches, readers can refer to [2, 3, 29, 30, 34, 43].

Moving from VQA to *Video Grounding*, we can see that *Video Grounding* is also a suitable application of NMN approaches, but it should be decomposed in a different manner from VQA. Currently, there are not many methods involving NMN. TMN [50] is a typical NMN approach, where query phrases are transformed into the compositional structure and then guide the assembling of modules. Figure 4.23 shows how modules are assembled. Temporal Compositional Modular Network (TCMN) is another example, which is visually illustrated in Fig. 4.24. This method composes multiple modules into one, first using a tree attention network to decompose a complex sentence into sub-descriptions and then using two modules for the measure of visual similarity and location similarity. Given the compositional nature of the method, however, it does not dynamically compose modules, making it less similar to typical NMN methods.

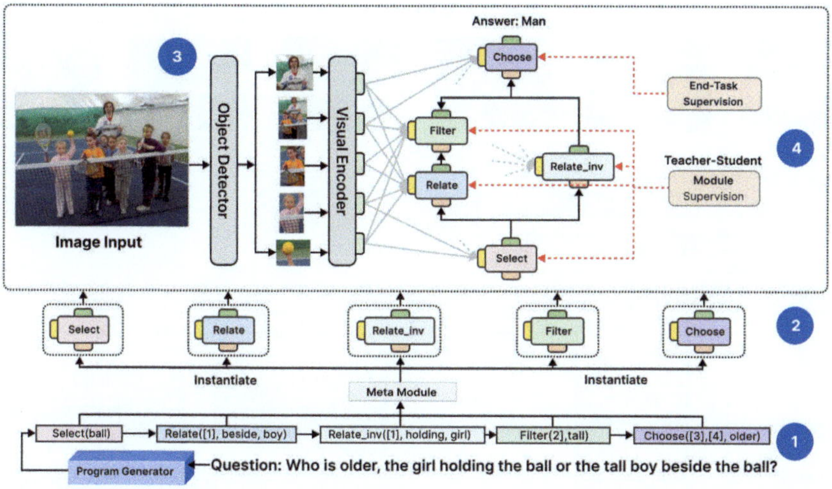

Fig. 4.22 MMN translates given questions into programs and instantiates programs as modules

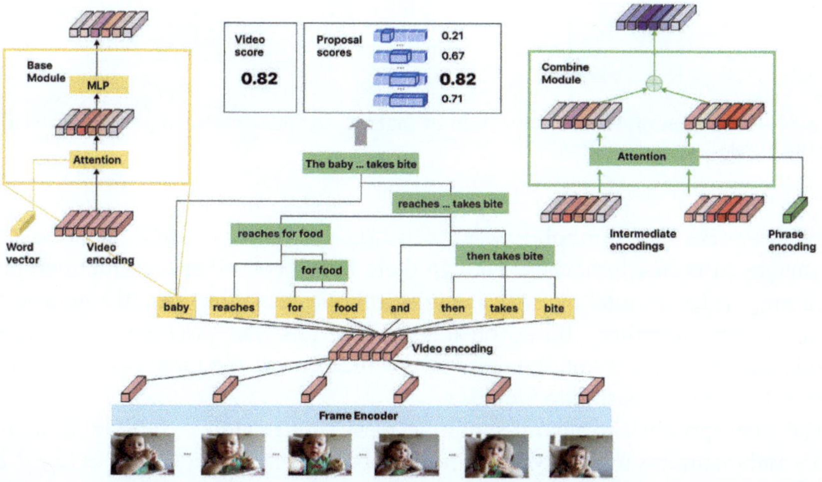

Fig. 4.23 The framework of TMN, where the compositional layout is not fixed, being dynamically assembled based on the given queries instead

4.2.3 *Grounding via Neural-Symbolic Reasoning*

Neural networks possess strong perception and learning abilities but are weak at reasoning, especially in terms of explainability. Compared to neural networks, a more conventional approach of symbolic systems shows better reasoning capability but is weaker at learning. Hence, targeting more explainable neural networks, a new approach naturally emerges: neural-symbolic reasoning [18, 87]. By integrating

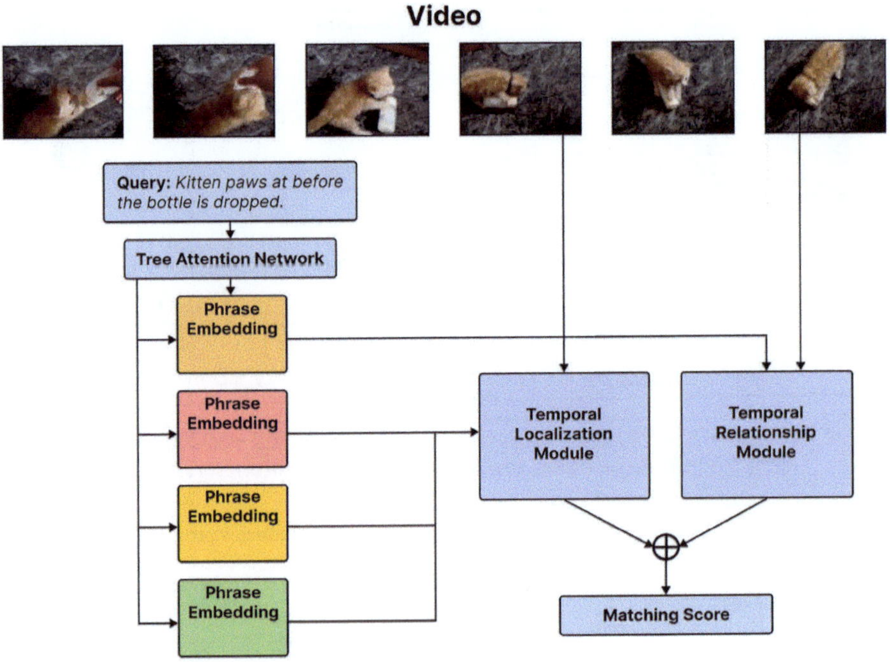

Fig. 4.24 Illustration of TCMN, comprised of multiple modules to decompose a sentence into multiple sub-descriptions

neural networks and symbolic systems, neural-symbolic reasoning possesses the advantages of both approaches. Though there have been no successful attempts at combining video grounding with neural-symbolic reasoning yet, we believe this is a promising direction. Based on this insight, this part provides a preliminary introduction to neural-symbolic reasoning, which covers the three major categories of neural-symbolic reasoning methods.

The first approach to neural-symbolic reasoning regards symbolic systems as basics and integrates neural systems into symbolic systems, as shown in Fig. 4.25a. Typical symbolic systems take structured data such as knowledge bases as input and solve problems by searching in the solution space. The added neural systems modify the framework of symbolic systems in two ways. On one hand, neural systems help transform unstructured data, which is more suitable for neural systems, into structured data which is processed by symbolic systems. On the other hand, neural systems promote the solution-searching process, shrinking the solution space that needs searching. Hence, the first approach enables symbolic systems to process unstructured data and promotes their efficiency. The pLogicNet model [68] utilizes neural networks to solve the inference problem on Markov logic networks. The problem was originally solved by inference techniques of graphical models, while pLogicNet uses neural networks and variational expectation-maximization to approximate the inference process. Neuro-Symbolic Concept Learner (NS-CL) [61]

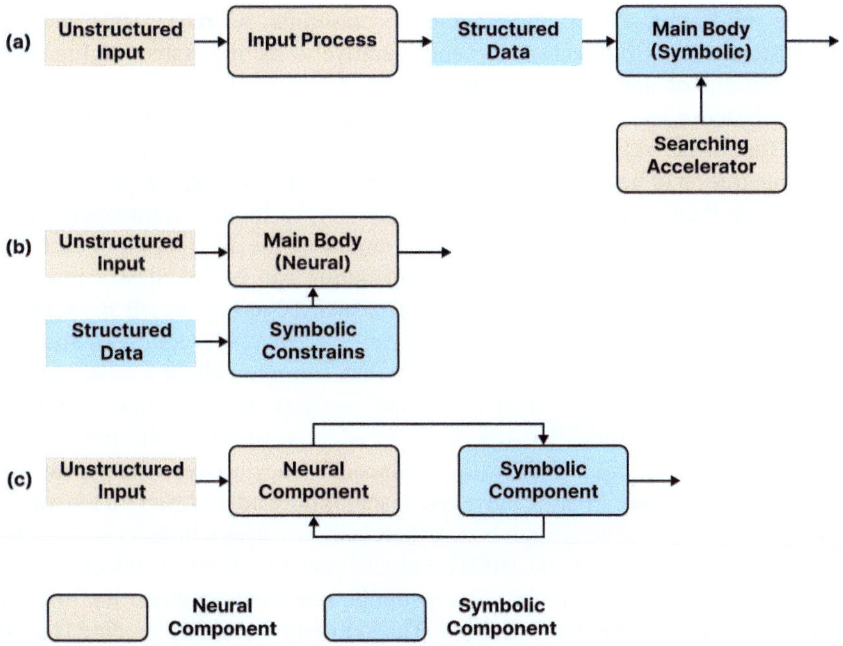

Fig. 4.25 Visual illustration of three major approaches to neural-symbolic reasoning

is a model for visual question answering that converts vision and text information into symbolic representation. NS-CL uses a visual perception module to extract object representation from a scene, a semantic parsing module to convert a question into an executable program, and a symbolic reasoning module to infer the program on the scene.

The second approach to neural-symbolic reasoning instead regards neural systems as basics and integrates symbolic systems into neural systems, as depicted in Fig. 4.25b. Typically, neural systems automatically learn from data, which is efficient and effective but hardly explainable. The introduced symbolic systems act as additional constraints to neural systems, such as input data, regularization terms, or training schedulers. Since symbolic systems are more explainable, the training procedure of neural systems has also become explainable. For instance, HDNN [28] utilizes knowledge distillation along with neural-symbolic reasoning. A teacher network learns from both labeled data and logic rules and then teaches a student network, thus constraining the training of the student model with symbolic systems. Another example is SEKB-ZSL, which conducts the image classification task. SEKB-ZSL [77] first trains an image classifier on visual data and a node classifier on knowledge graph embeddings, which also include unseen classes with regard to visual data. Afterwards, the weights of the node classifier are utilized to supervise the training of the image classifier, transferring knowledge from knowledge graphs to neural networks, enabling the model to correctly classify unseen images.

In the third approach to neural-symbolic reasoning, neural systems and symbolic systems work bidirectionally and iteratively, as illustrated in Fig. 4.25c. The predictions of neural systems are projected into structured data as inputs to symbolic systems, and the solutions found by symbolic systems are input into neural systems for further refinement and iteration. Such an alternating process enables deeper integration of neural and symbolic systems. For example, DeepProbLog [60] combines probabilistic logic with neural networks. In this framework, neural networks are utilized as an image classifier, outputting the distribution of each class in a dataset. Afterwards, the predicted distribution can be seamlessly integrated into a probabilistic logic inference system, which is mostly based on symbolic systems. A different approach, abductive learning (ABL) [93], combines abductive reasoning with abduction mutually. Given a data sample without a label, a neural network generates a pseudo label for it and converts the label into a symbolic representation. Then, abductive reasoning is adopted to revise the pseudo label based on a knowledge base, minimizing the inconsistency between the pseudo label and the knowledge base. Finally, the revised label is utilized to supervise the training of the neural network. Another method, BPGR [86], involves a bi-level probabilistic graphical reasoning framework. BPGR extracts visual features of objects from an image with a neural network and uses symbolic reasoning to guide the inference of visual concepts as an error correction. The symbolic reasoning module is a two-layer probabilistic graph consisting of top nodes as the reasoning results of the visual reasoning module and bottom-nodes logic rules.

The aforementioned successes of neural-symbolic reasoning indicate that video grounding benefits greatly from neural-symbolic reasoning. Besides better explainability, there also exist other benefits: (i) The symbolic inference process may grant the model stronger overall capability, achieving better performance. (ii) Some neural-symbolic reasoning methods introduce external knowledge bases for symbolic inference, which brings the possibility of knowledge-based or knowledge-driven video grounding. (iii) The symbolic inference component requires an explicit reasoning process, thus preventing the model from taking shortcuts, which is common in *Video Grounding* datasets.

Though video grounding has not been associated with neural-symbolic reasoning, there have been successful examples for other related tasks, which might imply the feasibility of solving *Video Grounding* with this method. For instance, Logic Tensor Network (LTN) [13] is a method for image interpretation with regard to object relations. It can effectively reason from noisy images while describing image characteristics through logic rules. MAC [32] (Fig. 4.26) is another method for visual question answering, decomposing images and questions into sequential units, and iteratively reasoning with strong prior conditions. This demonstrates that neural-symbolic reasoning can achieve a basic understanding of visual content. Sikka et al. [73] proposes a method for few-shot classification as shown in Fig. 4.27, which utilizes external knowledge bases to promote generalization ability. This success indicates that neural-symbolic reasoning might be a good solution for video

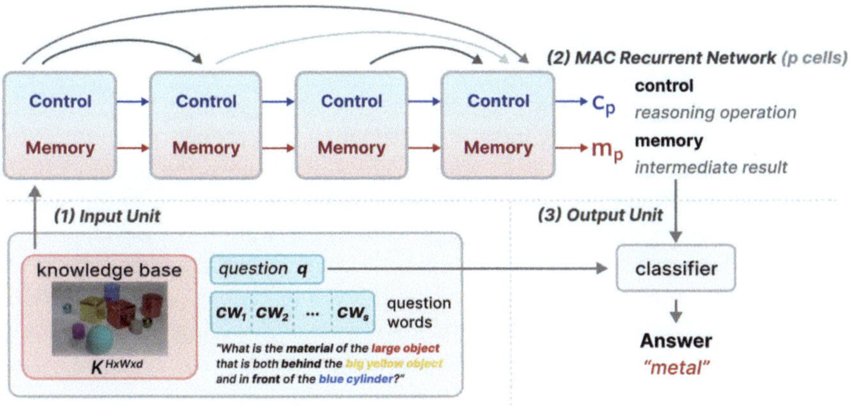

Fig. 4.26 The framework of MAC, a fully differentiable network with cyclic memory, attention, and composition functions

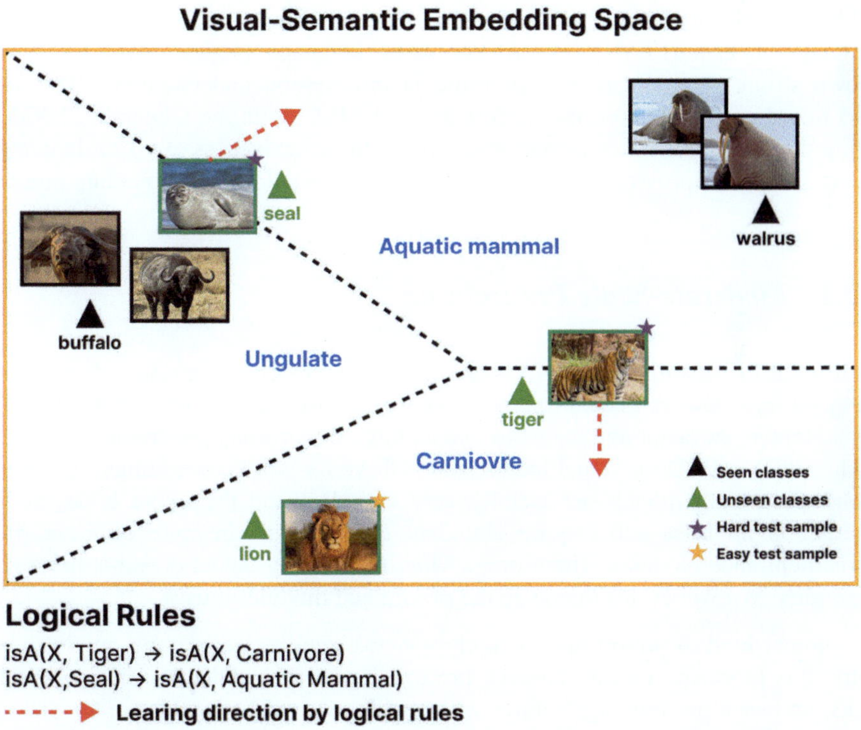

Fig. 4.27 The framework of [73], which regularizes the loss function according to logic rules

grounding in O.O.D. scenarios. Considering the relation between video grounding and the tasks described above, solving *Video Grounding* with neural-symbolic reasoning seems feasible.

In summary, neural-symbolic computation encompasses high-level integration of cognitive abilities (including induction, deduction, and abduction) and the study of how brains make mental models, among others, while covering the modeling of emotions and attention/utility. At a computational level, it addresses the study of the integration of logic, probabilities, and learning, as well as the development of new models of computation combining robust learning and efficient reasoning. With regard to applications, successes have been achieved in various domains, including simulation, bioinformatics, fault diagnosis, software engineering, model checking, visual information processing, and fraud prevention.

4.3 Vision-Language Pre-training for Generalized Video Grounding

With the rapid development of large-scale vision-language pre-training for large language models (LLMs), multimodal large language models (MLLMs) have shown strong generalization in multimedia information understanding. This has also raised concerns about the performance of MLLMs in the Generalized Video Grounding tasks. Numerous experiments prove the effectiveness of vision-language pre-training in improving the performance of generalized Video Grounding tasks.

4.3.1 Moderate-Scale Pre-training

Unimodal Pre-training Versus Vision-Language Pre-training *Video Grounding* methods have always been strongly related to pre-training. More specifically, the visual feature extractor and the word embeddings are typically pre-trained modules, such as I3D or C3D for visual features and GloVe for word embeddings. However, *Video Grounding* models are usually very complex, and their core cross-modal interaction modules still require abundant training data in spite of these pre-trained entrance modules. Therefore, a *Video Grounding* model does not naturally generalize to novel events thanks to the pre-trained modules it uses.

Another kind of pre-training, namely vision-language pre-training, might overcome this problem. Vision-language pre-training is designed for vision-language tasks, and such pre-training features are capable of providing guidance for vision-language alignment. In this section, we would like to emphasize vision-language pre-training and introduce how they can help *Video Grounding* models reach better performance.

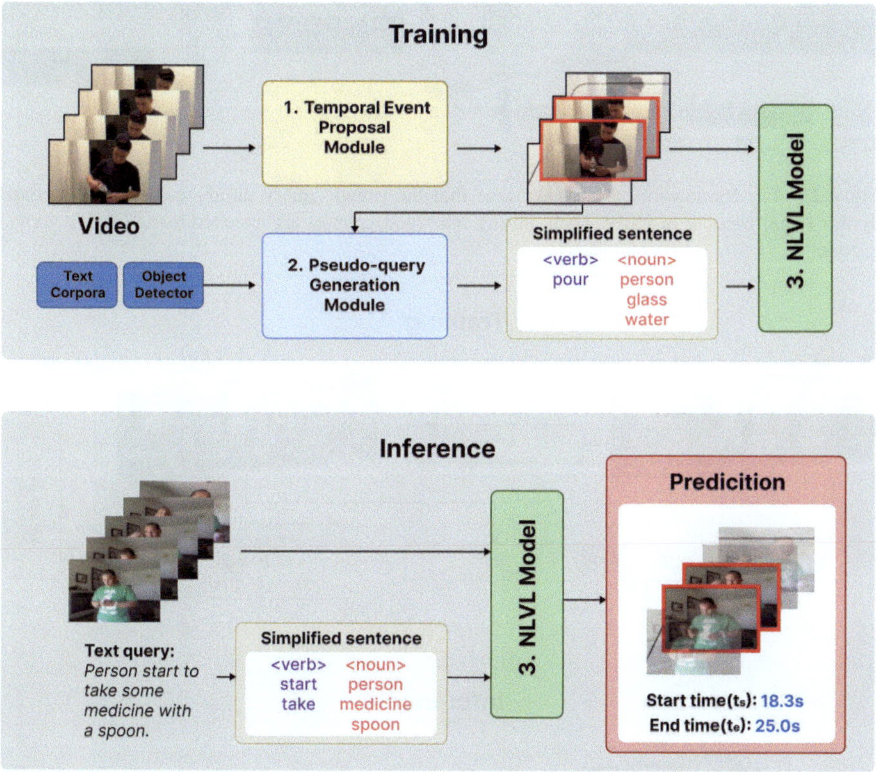

Fig. 4.28 Overview of PSVL framework. During training, Temporal Event Proposal Module finds moments of interest in the whole video, and Pseudo-Query Generation Module generates pseudo queries consisting of nouns and verbs. Through inference, the given query is broken into nouns and verbs to be compatible with the training schema

There are currently two types of research on this topic. One aims to make use of public vision-language pre-training models such as CLIP, while the other aims to design a new pre-training framework.

Nam et al. [63] are the first to define the zero-shot *Video Grounding* task, i.e., only a set of videos are available, while neither corresponding queries nor moment annotations are provided. To address this problem, they propose to first find meaningful temporal regions via clustering visual information and then use an off-the-shelf object detector along with noun-verb statistical co-occurrence patterns to generate pseudo queries, as shown in Fig. 4.28. With pseudo queries, it is possible to adapt any fully supervised *Video Grounding* model to the zero-shot setting. However, since the generated pseudo queries are just collections of words instead of structured sentences, the potential of this method is heavily limited.

The following work [37] bypasses the problem of generating structured pseudo queries with the help of CLIP [71], as shown in Fig. 4.29. Since CLIP aligns

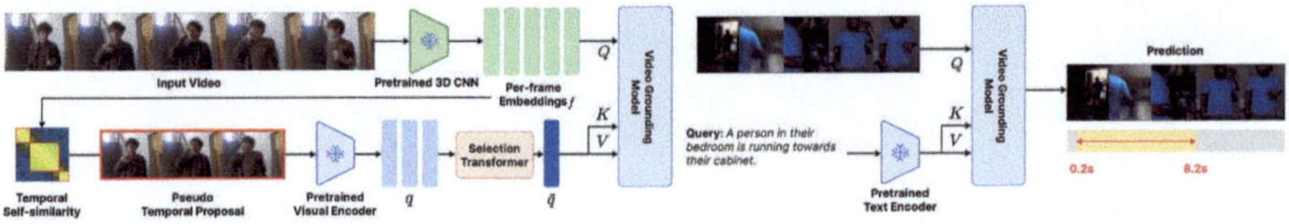

Fig. 4.29 The framework of [37]. We note that the pseudo query during training is generated by the visual encoder of CLIP. Then, during inference, queries are encoded using CLIP's textual encoder

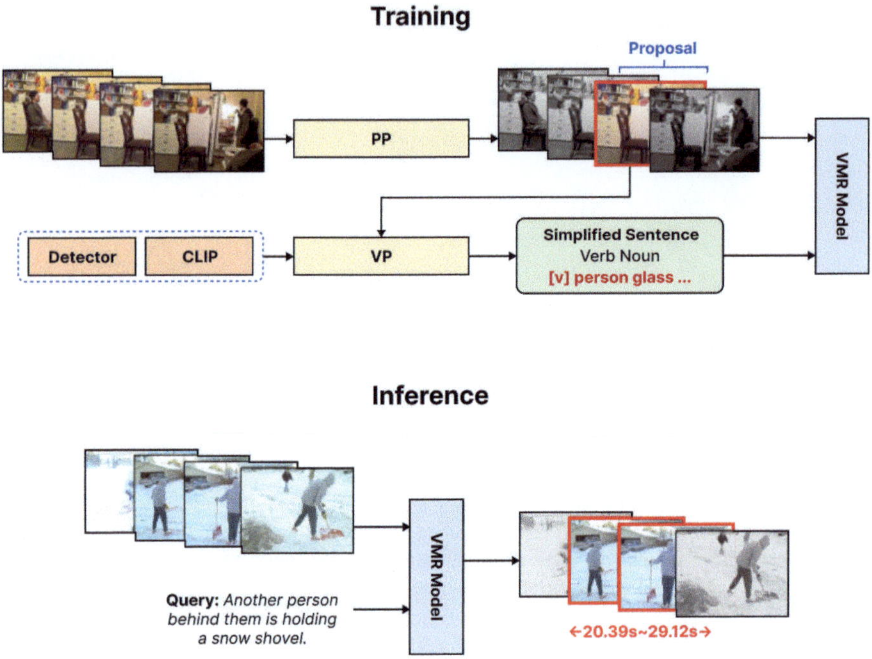

Fig. 4.30 Overview of the PZVMR framework. Proposal Prompt (PP) is designed to help find moments of interest, and Verb Prompt (VP) is designed to generate better pseudo queries

corresponding visual and textual features together, Kim et al. simply use the dominant visual feature extracted from a video moment as the pseudo sentence feature. This method outperforms the previous method [63], demonstrating its effectiveness.

Wang et al. [78] propose an alternative way of improving the performance of [63]. Inspired by the schema of prompt learning, they designed Proposal Prompt that masks sequential frames and Verb Prompt that masks action words. They also utilize CLIP to recognize proposal moments as well as verbs from the video and match queries with them to generate pseudo queries with better quality (Fig. 4.30).

Both [37] and [78] introduce CLIP to the zero-shot *Video Grounding* setting and achieve better performance with its help. It should be noted, however, that all three methods above show much worse results than state-of-the-art fully supervised models, as they are all under the zero-shot setting. Moreover, though CLIP is a vision-language pre-trained model with the ability of vision-language alignment, it cannot capture temporal information, thus restricting its application in *Video Grounding* research.

While the aforementioned methods simply utilize pre-trained models, the method proposed by Luo et al. [57] performs pre-training and achieves much better results. Although this work [57] has been explained in Sect. 4.1.3, Luo et al.'s work [57] is, in essence, an attempt to transfer CLIP to video tasks, or a "secondary pre-training" of a pre-training model. Since CLIP cannot capture temporal information, an additional model with access to CLIP is trained on video tasks to capture video changes and actions.

In conclusion, vision-language pre-training models are shown to be beneficial to *Video Grounding* methods, but simply integrating CLIP into a *Video Grounding* model results in poor temporal perception. The solution, however, requires a secondary pre-training procedure with abundant training data to transfer CLIP to video tasks. Although utilizing vision-language pre-training models is a promising approach, current methods are still far from satisfactory. While it is safe to continue the attempt to find a better approach to integrating conventional vision-language pre-training models, we will present a more aggressive direction next, to which we hope more attention could be drawn.

Hypothesized Improvement to Existing Methods Having discussed previous works, we notice that a major challenge of applying vision-language pre-trained models to the *Video Grounding* task is that most of them are actually image-language models. Such models lack the necessary ability to capture temporal information and thus require additional techniques for transferring. Video-language models, however, are much rarer and have not been explored in *Video Grounding* research yet.

How can we overcome the gap between image and video? We notice that the most prevalent big models nowadays, i.e., LLM and diffusion models, have not been exploited in *Video Grounding* task yet. This seems to be a promising approach, as both GPT-4 and diffusion models have cross-modal capability, and they reach much more impressive performance than previous pre-trained models.

More specifically, although GPT-4, as a multimodal LLM model, can process only images but not videos, there are some further studies following its success. For example, Wang et al. [80] build ChatVideo, a generic video understanding system, as a combination of various video foundation models and LLM. They first parse the video via video foundation models and then pass the resulting information to LLM to provide natural language interaction ability. This work shows very impressive results, yet it is still a multi-step approach, limiting the system's capability of directly understanding videos.

Among diffusion models, the most common type is text-to-image diffusion models such as [72], which do not have the ability for video understanding. Moreover, there has been some research on video diffusion models. Starting with Ho et al.'s work [26], many researchers have devoted themselves to video diffusion models [24, 36, 59, 94]. However, as diffusion models are generative models, and do not have the ability of natural language interaction as LLM, it is difficult to adapt off-the-shelf video diffusion models to the *Video Grounding* task.

To conclude, the rapid development of big foundation models is offering more and more handy tools for various downstream tasks, but how to adjust them to accomplish the *Video Grounding* task is non-trivial. Therefore, the major target of future *Video Grounding* research should be finding a way to adapt big foundation models to the *Video Grounding* task.

4.3.2 Emergent LLMs and Multimodal LLMs

Pioneered by the famous ChatGPT, large language models lead to a new era of artificial intelligence. As shown by the success of moderate-scale pre-training, large base models can bring about even more significant improvement to both the methodology and performance of video grounding. In this part, we aim to first provide the necessary background information on LLMs and MLLMs.

LLMs are large base models for natural language processing. The mainstream LLMs are based on transformer architectures, such as BERT [12] and GPT [85]. Specifically, the decoder-only architecture of GPT is especially beneficial for scaling up, thus having become the prevalent architecture nowadays. As the scale goes up, intelligence emerges within LLMs, giving them impressive conversational capability. Besides ChatGPT, which is close-source, there are also open-source LLMs. For example, ChatGLM and its newer version GLM-4 [20], OpenLLaMA [19], and LLaMA 2 [76] are all LLMs with both strong capability and open codes. However, LLMs are restricted to language-only tasks, so the attempt to extend LLMs to vision tasks leads to MLLMs.

Multimodal large language models, as indicated by the name, are LLMs further integrated with multiple modalities rather than independent models. As a result, both tasks for MLLMs and their implementation methods are based on LLMs. The major task for MLLMs, similarly to LLMs, is performing dialogues. Instead of using only dialogues in textual modality, MLLMs enable users to input images or even videos, which can be better perceived and understood by the model to engage in the conversation. However, it should be noted that MLLMs are not expected to generate visual outputs; they only generate textual responses. Method-wise, MLLMs project both textual and visual prompts into a common embedding space using dedicated encoders, which is similar to how vision transformers [14] encode images. Afterwards, the projected embeddings are processed by the stacked transformers in a similar way to text-only LLMs. Examples of MLLMs include relatively smaller ones such as Minigpt-4 [95], Kosmos-2 [67], and DeepSeek-VL [55], and larger ones such as GPT-4 [1] and LLaVA 1.6 [54]. Specifically, LLaVA 1.6 achieves high image resolution by chunking images into blocks to encode separately and downsample the original image for global perception.

Having clarified the basic concept of LLMs and MLLMs, we move on to discuss how they are pre-trained and then applied to downstream tasks. For pre-training, large base models also require large-scale training data, which is usually web data. Such web data is not delicately processed but remains in a rather raw state, which poses restrictions to the pre-training task. To be specific, the pre-training task should not be too complicated nor too specific, which may only suit training samples of a certain format. In contrast, the task should be in a very simple form, not requiring any specific format or structure. Consequently, the common practice is to predict masked tokens and the next tokens, as visually illustrated in Fig. 4.31. Between the two, decoder-only architectures prefer next token prediction, which feeds the model

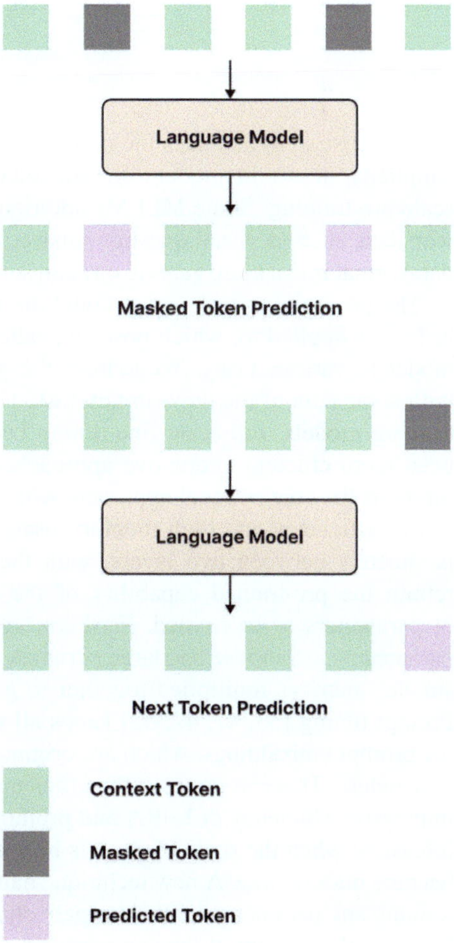

Fig. 4.31 Illustration of two common pre-training tasks for LLMs. Masked token prediction randomly masks some tokens in the input and asks the model to predict the masked tokens, while next token prediction allows the model to predict the next token given some prefix tokens

Fig. 4.32 Four types of LLM tuning, with a decreasing number of tuning parameters

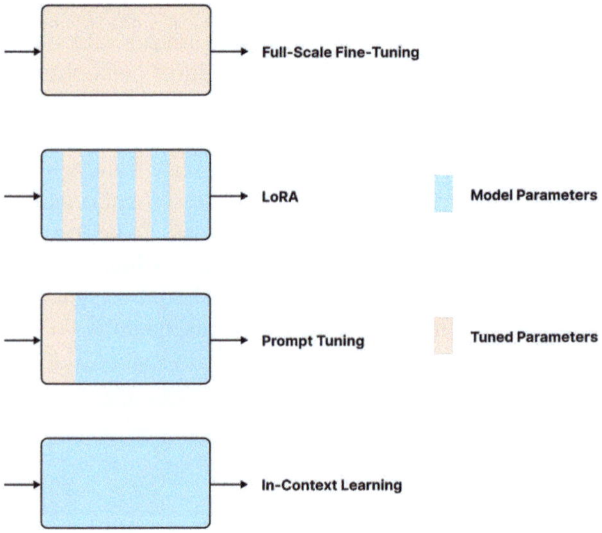

with prefix tokens and asks the model to correctly output the next token [70]. Such simplicity allows the model to be trained on various web data, thus enabling large-scale pre-training. Some MLLMs additionally train on benchmark datasets besides web data, such as visual question-answering datasets. In such cases, modifying the data format into a more generic one can sometimes benefit the model capability [54].

The pre-training tasks of a model are not usually the same as the downstream tasks it is applied to, which poses a challenge if we wish to adapt one pre-training model to various tasks. To address the gap, the common practice is fine-tuning with some data of the downstream task. However, with the increasing scale of pre-training models, full-scale fine-tuning becomes harder. Consequently, there have been more efficient alternative approaches, as shown in Fig. 4.32. Full-scale fine-tuning is the conventional approach, which trains all the model parameters, leading to low efficiency and high memory usage. LoRA [31] inserts additional trainable parameters between two layers, with the original parameters frozen. This helps retrain the pre-trained capability of the original model and reduce the number of parameters to be trained. Furthermore, as illustrated in Fig. 4.33, the inserted parameters decompose one large parameter matrix into two low-rank and thus much smaller matrices multiplied together to greatly reduce the number of parameters. Prompt tuning [33, 41, 42, 52] keeps all the original parameters frozen and treats the prompt embeddings, which are originally calculated by an encoder, as trainable parameters. This can improve the efficiency of fine-tuning even further. Despite the impressive efficiency of LoRA and prompt tuning, the two techniques also become infeasible when the scale of models becomes larger or when the model parameters become inaccessible. A new technique named in-context learning thus emerges, not training any parameter at all. Alternatively, LLMs have the capability of inferencing based on the context of the conversation. Given such capability, in-context learning

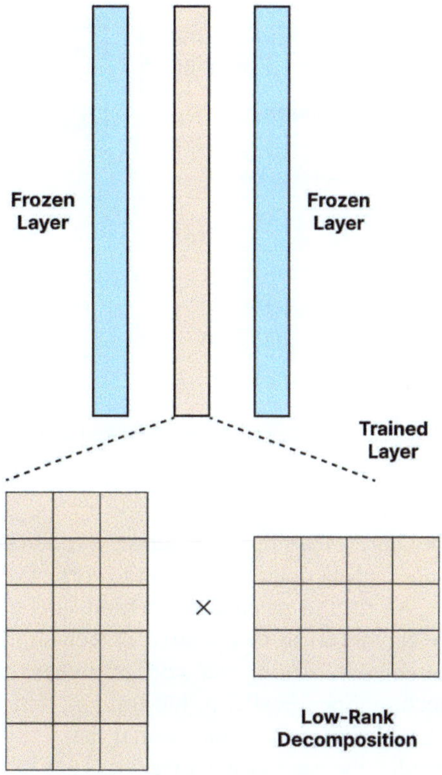

Fig. 4.33 LoRA inserts trainable parameters between frozen layers and decomposes a large matrix into two low-rank matrices

provides a few examples consisting of questions and answers in a certain format within the context, and LLMs can learn from the context how to answer similar questions. Given its effectiveness and efficiency, in-context learning has become a highly prevalent technique for LLMs and MLLMs.

In-context learning is currently a very important tool for research on LLMs and MLLMs, so we provide a more detailed explanation here. As a typical usage of in-context learning, we need to provide the model with several examples and one query, all in the same format. As shown in Fig. 4.34, the model infers from the context to understand the correct rules for answering the given query. Formally, this process can be described as giving the model a demonstration set of $C = I, s(x_1, y_1), \cdots, s(x_k, y_k)$, where I is the instruction in natural language and $s(x_i, y_i)$ is one demonstration example. The likelihood of one candidate for the answer is a score function conditioned on the whole input sequence:

$$P(y_j|x) \triangleq f_M(y_j, C, x), \tag{4.5}$$

and the final prediction is the candidate with the highest score:

$$\hat{y} = arg \max_{y_j \in Y} P(y_j|x). \tag{4.6}$$

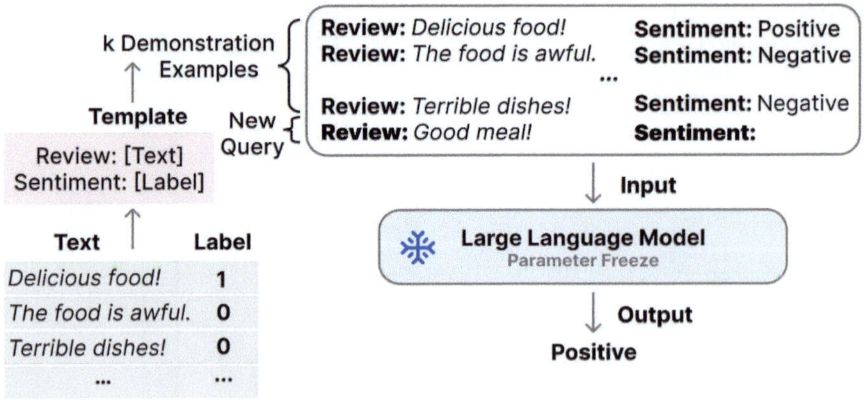

Fig. 4.34 In-context learning provides LLMs with several examples and one query, all in the same format

In-context learning, in essence, is modifying the context C such that the conditional scoring function changes in the correct direction.

As for the mechanism behind in-context learning, it is a unique and emergent phenomenon in recent advancements in LLMs. The current explanation as to why in-context learning can work attributes this ability to both data and the model architecture. Specifically, in the training data for LLMs, many samples involve several statements of the same format listed together. Therefore, the model can correctly predict the next tokens more easily if it pays attention to previous examples, so the model is encouraged by its training data to perform in-context learning. On the other hand, the attention mechanics vastly used in LLM architectures are also beneficial to in-context learning, as they enable the model to especially attend to the same relative location of multiple examples, making it easier to follow the given format. A simple example to illustrate the mechanics behind in-context learning involves two-layer attention structures and some toy data, as shown in Fig. 4.35. To be specific, one $A_i - B_i$ pair works as a complete example, where A_i is the query and B_i is the answer. The context contains several example pairs and one query. The first attention layer aggregates the information from the current position (B2) and the previous position (A2). Afterwards, in the second attention layer, the information from the previous location (A2) acts as the key and query, and the information from the current position (B2) acts as value. In this way, the masked token matches the correct example (A2–B2) in context by query-key similarity computation and takes the value (B2) as the output. Though this example is simplified, similar mechanisms are believed to exist in actual LLMs, which explain the ability of in-context learning: the model learns from specific training data to look up the context and find answers instead of remembering the answers.

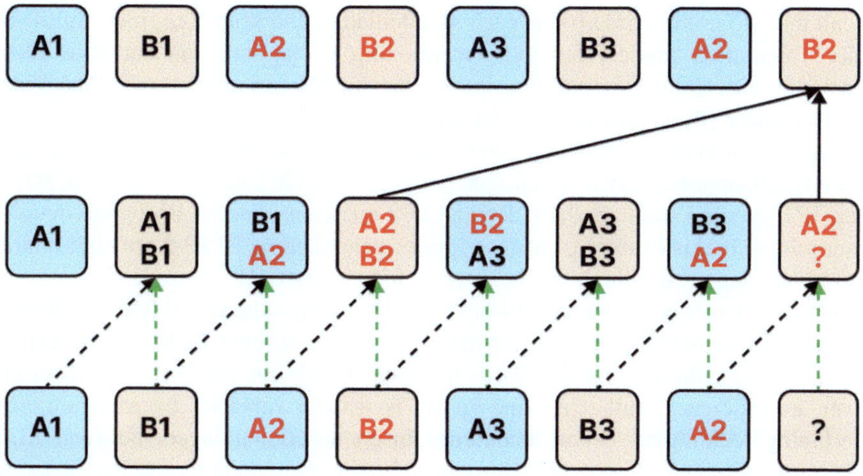

Fig. 4.35 A two-layer attention structure and specially constructed toy data can achieve the in-context learning capability

Fig. 4.36 Vid-LLM can be constructed in three approaches, i.e., (i) Using video models as external tools to generate textual descriptions; (ii) Using video models as external tools but directly rely on video embeddings; (iii) Pre-training a video-specific model from scratch that can naturally understand videos

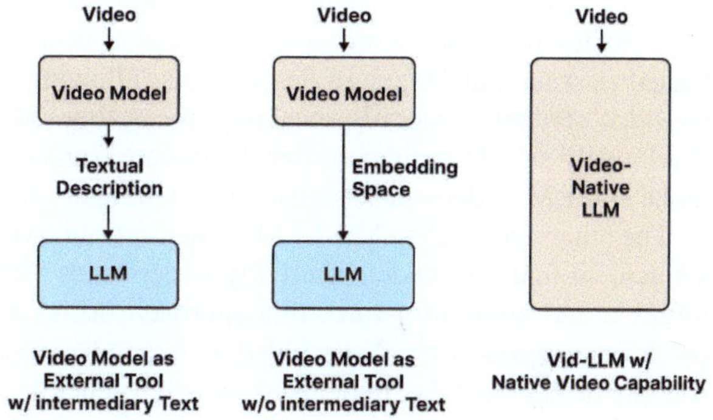

4.3.3 *LLMs for Videos*

The last part introduces both text-only LLMs and MLLMs, which are additionally integrated with the vision modality. However, note that the vision modality in MLLMs is static images instead of videos unless explicitly stated otherwise. Hence, MLLMs may fail to be directly applied to video tasks. Given this limitation, there have also been attempts on video-specific LLMs, which can be denoted as Vid-LLMs [75]. Three major approaches to Vid-LLMs are illustrated in Fig. 4.36 and will be discussed in detail next.

The first approach is to combine video foundation models with text- or image-only LLMs by letting LLMs use video models as tools. Some LLMs have the ability to call the APIs of external tools, such as calculators and search engines. Similarly, video foundation models for video captioning or tagging can serve as tools, translating videos into textual descriptions that LLMs can understand. Take Video ChatCaptioner [9] as an example, the key to this approach is to generate detailed descriptions of video content that can be directly understood by LLMs. Specifically, Video ChatCaptioner forms a dialogue between ChatGPT and BLIP-2 [46], where ChatGPT selects a frame in the video and asks questions, and BLIP-2 gives detailed descriptions. In this example, the descriptions generated by BLIP-2 are high-quality natural language text, which can be directly fed to ChatGPT as inputs.

However, cases occur when we want LLMs to process the visual features directly without an intermediary natural language modality. To help LLMs better exploit visual features, there are methods to fine-tune LLMs on video-specific datasets. Given its efficiency, full-scale fine-tuning is usually replaced by more efficient fine-tuning techniques. Some techniques insert additional layers between LLM parameters, similar to LoRA [31] introduced before. Other techniques may insert an additional layer between the video foundation model and the LLM, which somehow resembles the prompt tuning technique [33, 41, 42, 52]. For example, VALLEY [58] encodes video frames with CLIP and aggregates frame features using a temporal module, forming video features. Afterwards, these video features are projected into the textual embedding space through a trainable adapter layer rather than turned into natural language descriptions, allowing LLMs to work more directly on video content, Generally speaking, integrating video models as tools into LLMs is relatively simple and can utilize the mature fruits of LLMs. However, it is hard to make an LLM understand video content inherently through this approach.

The other approach to Vid-LLMs is to pre-train a video-specific large model from scratch, so that the model inherently understands video content. While seemingly intuitive and straightforward, this approach brings a new challenge of designing pre-training tasks for videos. LaViLa [92] utilizes video-narration pairs as a pre-training task, where the model is expected to generate accurate and diverse narration given a video. In addition, to enable the process of videos, LaViLa extends the LLM architecture with cross-attention modules that attend text to visual information. VAST [10] further adds more diverse pre-training tasks, using omni-modality video-caption contrastive, matching, and generation losses, which cover video question answering and video retrieval tasks beyond video captioning. Merlin [88] handles more versatile video understanding tasks, such as object tracking, video referring, video relation, and future reasoning, as pre-training tasks. In conclusion, pre-training a video-specific Vid-LLM from scratch on video tasks grants models the native video understanding ability, which has the potential to outperform the first approach.

If we aim to adapt Vid-LLMs to a downstream task, the efficient fine-tuning techniques introduced before can still be applied. LoRA and prompt tuning work as efficient fine-tuning with a small amount of trainable parameters, while in-context learning totally alleviates the need for further training. The fine-tuning techniques, along with the inherent capability of pre-trained Vid-LLMs, imply strong performance on various vision-language tasks.

4.3.4 Video Grounding as Pre-training Task for Vid-LLMs

The combination of video grounding and Vid-LLMs comes in two forms: *Video Grounding* can possibly serve as either a pre-training task or a downstream task. In this part, we discuss the possibility of utilizing *Video Grounding* as a pre-training task for Vid-LLMs.

Strict-Form *Video Grounding* Task Benchmark datasets of *Video Grounding* task can be converted into a strict-form pre-training task. Specifically, one data sample in such a dataset consists of a video, a sentence, and a timestamp, which can be transformed into a natural language sentence suitable for Vid-LLM pre-training. In this way, *Video Grounding* benchmarks can serve as a source of high-quality training data on a relatively small scale. Compared to large-scale web data, which may be of lower quality, *Video Grounding* benchmark data can play a complementary role.

Free-Form *Video Grounding* Task In large-scale web data, some may naturally reflect the demand from video grounding. To be specific, on video websites, videos come with such timestamps and corresponding subtitles that users may use to skip to intended parts of the videos. Additionally, in video descriptions and comments, users may use timestamps to refer to a time interval in the video, paired with descriptive or critical text. Examples of the two types of free-form timestamps are shown in Fig. 4.37. In both cases, such web data demand similar video understanding capability of Vid-LLMs to *Video Grounding* datasets.

Given both the strict-form and free-form *Video Grounding* tasks as part of the pre-training task, it is possible to equip Vid-LLM with the ability to perform video grounding. Video grounding is a fine-grained task compared to action recognition and video captioning, and the task requires the model to correctly establish the correlation between video content and textual descriptions. Such requirements may also benefit other video understanding tasks, leading to a generically capable large model.

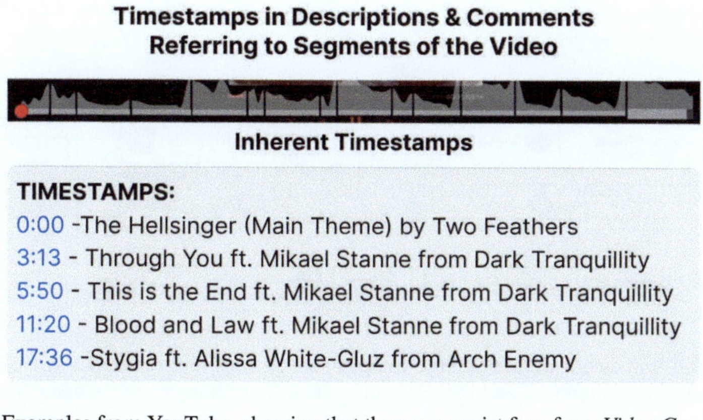

Fig. 4.37 Examples from YouTube, showing that there may exist free-form *Video Grounding* data in large-scale web data

4.3.5 *Video Grounding as Downstream Task for Vid-LLMs*

A Vid-LLM with generic video understanding capability may fulfill the requirement of video grounding, even without pre-training on the task. In this case, we can regard *Video Grounding* as a downstream task to which a pre-trained Vid-LLM should be adapted. Generally, such practice can follow the efficient fine-tuning methods for LLMs.

LLM4VG [16] is an early study of this direction, as illustrated in Fig. 4.38. It first builds a benchmark for testing Vid-LLMs against the *Video Grounding* task and then examines the performance of several models. Specifically, to adapt Vid-LLMs to *Video Grounding*, LLM4VG mainly designs specific prompts that guide Vid-LLMs to perform temporal grounding. Through prompting, LLM4VG asks the model to find the start and end timestamps of a described action. This work also examines the capability of text-only LLMs besides Vid-LLMs. For LLMs, LLM4VG first turns a model into a Vid-LLM by describing video content and providing an example in the dialogue context. Finally, LLM4VG concludes that current Vid-LLMs and LLMs show the preliminary capability of performing video grounding, but there is still room for improvement.

Grounding-Prompter [8] is another example, which is shown in Fig. 4.39. In this study, a Vid-LLM is first constructed by translating video content into textual

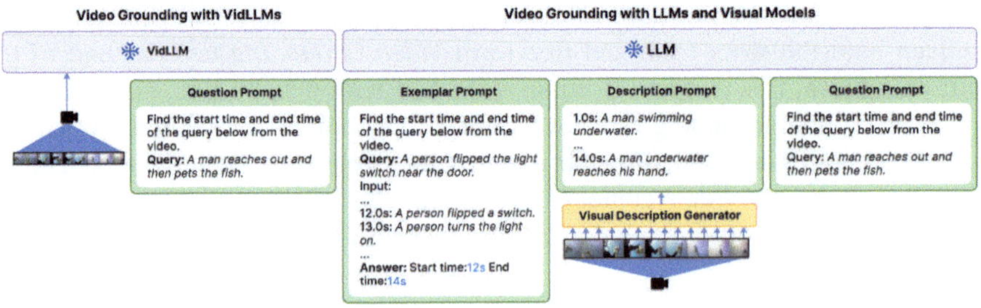

Fig. 4.38 The framework of LLM4VG, which designs specific prompts to guide LLMs and Vid-LLMs to perform video grounding

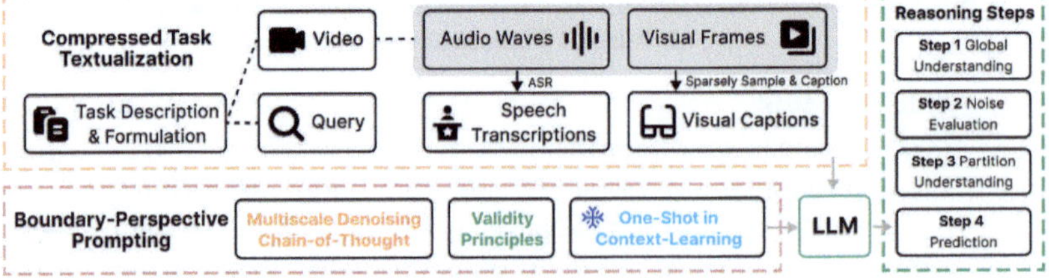

Fig. 4.39 The framework of Grounding-Prompter. which constructs a Vid-LLM based on text-only LLMs and video models, and then uses three delicate techniques to enhance the grounding performance

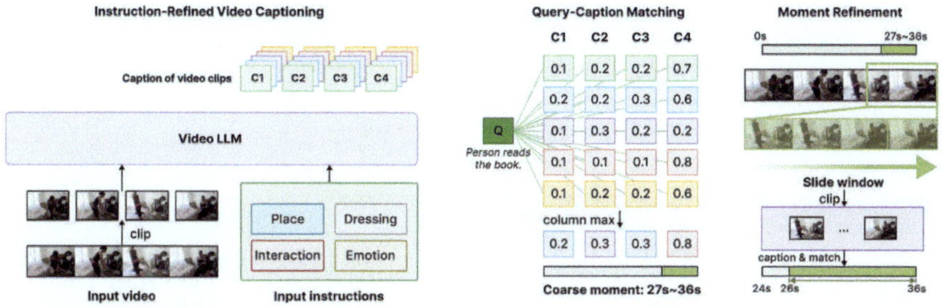

Fig. 4.40 The framework of ChatVTG, where the video grounding task is converted to multiple video captioning tasks, as well as the localization among textual descriptions

descriptions through video models. Afterwards, this Vid-LLM is adapted to *Video Grounding* in three ways: (i) A multi-scale Chain-of-Thought technique is utilized along with noise filtering to inspire the model to reason over local and global semantics. (ii) The model is asked to predict following a specific format, which puts additional restraints so that the model can generate reasonable predictions. (iii) In-context learning is integrated as an efficient fine-tuning replacement technique, giving an example for the model to imitate. The three additional designs grant Grounding-Prompter stronger capability than the previous method, which only designs specific prompts to adapt Vid-LLM to *Video Grounding*, showing good performance, especially on long videos. However, in-context learning is still relatively weak compared with fine-tuning, and the Vid-LLM is not pre-trained with *Video Grounding* task, which both restrict further enhancement.

ChatVTG [69] is another method with delicate designs for enhancement, as visualized in Fig. 4.40. Instead of directly asking Vid-LLMs to find the correct timestamps, ChatVTG asks Vid-LLMs to generate captions for segments of various granularity. Afterwards, the localization of textual descriptions is no longer performed on videos but on these captions, which is now a uni-modal task. As captions generated in this way are generally coarse, ChatVTG also designs further refinement methods. The motivation behind ChatVTG is that as Vid-LLMs are pre-trained on video captioning tasks and *Video Grounding* is a novel task, Vid-LLMs might perform non-negligibly better on captioning than *Video Grounding*. Hence, by converting *Video Grounding* to multiple video captioning tasks, better performance can be achieved.

VTG-LLM [22] argues that the current insufficient *Video Grounding* capability of Vid-LLMs comes from the fact that they are not specifically tuned for the *Video Grounding* task on suitable datasets. To this end, the authors first construct a new dataset for instruction-tuning that covers video grounding and a few other tasks. Then a new model is proposed as illustrated in Fig. 4.41 to explicitly integrate timestamp information into Vid-LLMs. Besides the promising results, we argue that the contributed dataset can also serve as a good data source for not only fine-tuning but also pre-training, enabling inherently time-aware Vid-LLMs.

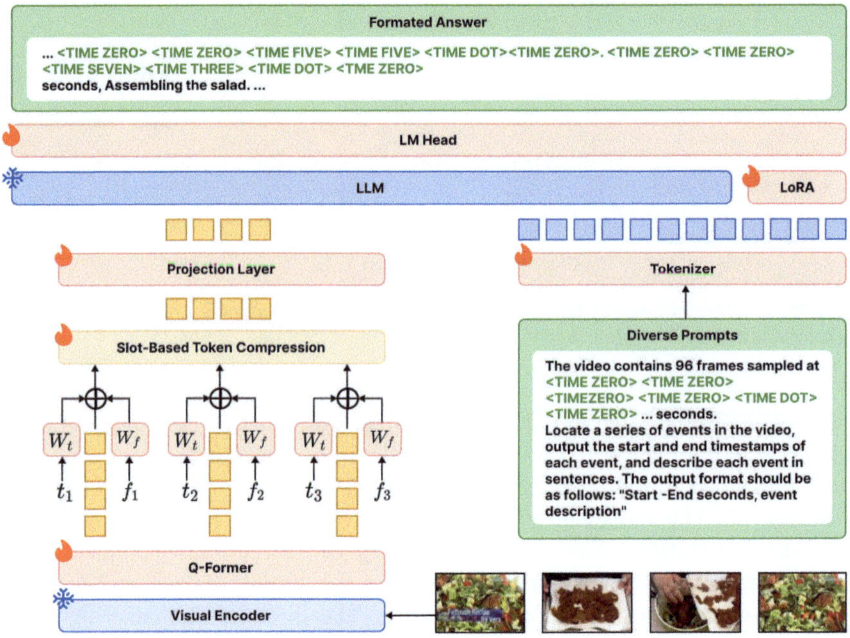

Fig. 4.41 The framework of VTG-LLM, which explicitly integrates timestamps into the model pipeline for better *Video Grounding* capability

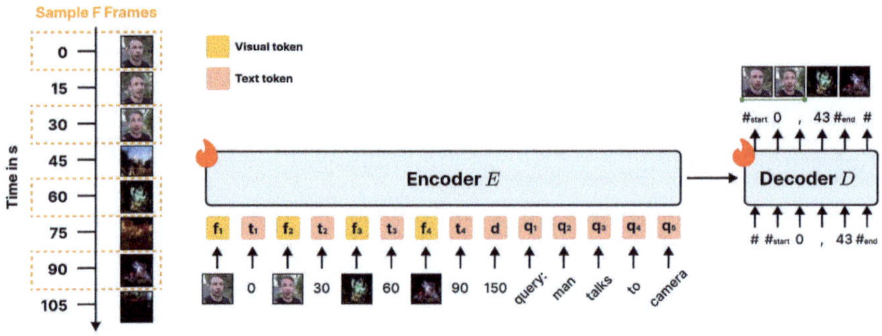

Fig. 4.42 Mr. BLIP fine-tunes BLIP-2 with the LoRA for the *Video Grounding* task

Mr. BLIP is another method that involves efficient fine-tuning instead of in-context learning, as shown in Fig. 4.42. It is based on BLIP-2, a rather small MLLM model, and further fine-tunes the model with LoRA on the *Video Grounding* task. Although the final result is inspiring, BLIP-2 is too small compared with more capable MLLMs and Vid-LLMs.

The five exploratory studies have shown promising potential for applying Vid-LLMs to video grounding, but there remains much to discover in future work. On

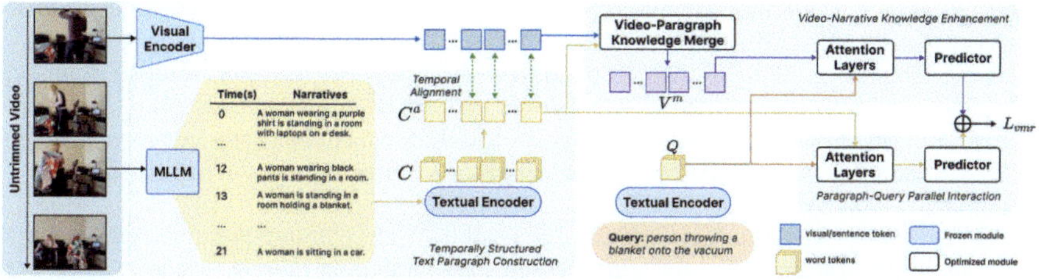

Fig. 4.43 The framework of a method that utilizes Vid-LLMs to generate annotations (narratives) for video segments

the one hand, it is interesting and worthy to study how to extend efficient fine-tuning such as LoRA and prompt tuning to larger Vid-LLMs, as better replacements for in-context learning. On the other hand, the *Video Grounding* task can serve as a valuable pre-training task for Vid-LLMs. Such a pre-training task can make a Vid-LLM natively come with the ability to perform video grounding, while generically promoting the video understanding capability of the model.

Cai et al. [6] propose an alternative approach to utilizing Vid-LLMs for *Video Grounding*: using Vid-LLMs to generate annotations for better diversity. Specifically, the authors argue that the current *Video Grounding* datasets face the problem of limited diversity with regard to annotations. The manual process of labeling has resulted in low wording diversity, which restricts the alignment ability of *Video Grounding* models to the limited word corpus. To this end, they propose to utilize Vid-LLMs to generate more diverse video descriptions, as illustrated in Fig. 4.43. Additionally, they have designed mechanics to retain good alignment of timestamps, including per-timestamp narrative, cross-modal feature merging between temporal-aware narratives and video temporal features, and an unimodal narrative-query matching mechanic.

Though different from the typical way to apply Vid-LLMs to downstream tasks, Vid-LLMs' capability to generate additional training data is also a good option. This is especially useful when the dataset is limited and annotation is expensive. However, the gap between generated data and real data should be addressed.

References

1. Achiam, J., Adler, S., Agarwal, S., Ahmad, L., Akkaya, I., Aleman, F.L., Almeida, D., Altenschmidt, J., Altman, S., Anadkat, S., et al.: Gpt-4 technical report. arXiv preprint arXiv:2303.08774 (2023)
2. Andreas, J., Rohrbach, M., Darrell, T., Klein, D.: Learning to compose neural networks for question answering. arXiv preprint arXiv:1601.01705 (2016)
3. Andreas, J., Rohrbach, M., Darrell, T., Klein, D.: Neural module networks. In: Proceedings of the IEEE Conference on Computer Vision and Pattern Recognition, pp. 39–48 (2016)
4. Auda, G., Kamel, M.: Modular neural networks: a survey. Int. J. Neural Syst. **9**(02), 129–151 (1999)

5. Bao, P., Mu, Y.: Learning sample importance for cross-scenario video temporal grounding. arXiv preprint arXiv:2201.02848 (2022)
6. Cai, W., Huang, J., Gong, S., Jin, H., Liu, Y.: Mllm as video narrator: mitigating modality imbalance in video moment retrieval. arXiv preprint arXiv:2406.17880 (2024)
7. Chen, W., Gan, Z., Li, L., Cheng, Y., Wang, W., Liu, J.: Meta module network for compositional visual reasoning. In: Proceedings of the IEEE/CVF Winter Conference on Applications of Computer Vision, pp. 655–664 (2021)
8. Chen, H., Wang, X., Chen, H., Song, Z., Jia, J., Zhu, W.: Grounding-prompter: prompting llm with multimodal information for temporal sentence grounding in long videos. arXiv preprint arXiv:2312.17117 (2023)
9. Chen, J., Zhu, D., Haydarov, K., Li, X., Elhoseiny, M.: Video chatcaptioner: towards enriched spatiotemporal descriptions. arXiv preprint arXiv:2304.04227 (2023)
10. Chen, S., Li, H., Wang, Q., Zhao, Z., Sun, M., Zhu, X., Liu, J.: Vast: a vision-audio-subtitle-text omni-modality foundation model and dataset. Adv. Neural Inf. Process. Syst. **36**, 72842–72866 (2024)
11. Cui, Y., Jia, M., Lin, T.Y., Song, Y., Belongie, S.: Class-balanced loss based on effective number of samples. In: Proceedings of the IEEE/CVF Conference on Computer Vision and Pattern Recognition, pp. 9268–9277 (2019)
12. Devlin, J., Chang, M.W., Lee, K., Toutanova, K.: Bert: pre-training of deep bidirectional transformers for language understanding. arXiv preprint arXiv:1810.04805 (2018)
13. Donadello, I., Serafini, L., Garcez, A.D.: Logic tensor networks for semantic image interpretation. arXiv preprint arXiv:1705.08968 (2017)
14. Dosovitskiy, A., Beyer, L., Kolesnikov, A., Weissenborn, D., Zhai, X., Unterthiner, T., Dehghani, M., Minderer, M., Heigold, G., Gelly, S., Uszkoreit, J., Houlsby, N.: An image is worth 16x16 words: transformers for image recognition at scale. In: 9th International Conference on Learning Representations (2021)
15. Fashandi, H.: Neural module networks: a review. Neurocomputing **552**, 126518 (2023)
16. Feng, W., Wang, X., Chen, H., Zhang, Z., Song, Z., Zhou, Y., Zhu, W.: Llm4vg: large language models evaluation for video grounding. arXiv preprint arXiv:2312.14206 (2023)
17. Ganin, Y., Ustinova, E., Ajakan, H., Germain, P., Larochelle, H., Laviolette, F., March, M., Lempitsky, V.: Domain-adversarial training of neural networks. J. Mach. Learn. Res. **17**(59), 1–35 (2016)
18. Garcez, A.d., Bader, S., Bowman, H., Lamb, L.C., de Penning, L., Illuminoo, B., Poon, H., Zaverucha, C.G.: Neural-symbolic learning and reasoning: a survey and interpretation. In: Neuro-Symbolic Artificial Intelligence: The State of the Art, vol. 342(1) (2022)
19. Geng, X., Liu, H.: Openllama: an open reproduction of llama (2023). https://github.com/openlm-research/open_llama
20. GLM, T., Zeng, A., Xu, B., Wang, B., Zhang, C., Yin, D., Rojas, D., Feng, G., Zhao, H., Lai, H., et al.: Chatglm: a family of large language models from glm-130b to glm-4 all tools. arXiv preprint arXiv:2406.12793 (2024)
21. Gorishniy, Y., Rubachev, I., Khrulkov, V., Babenko, A.: Revisiting deep learning models for tabular data. Adv. Neural Inf. Process. Syst. **34**, 18932–18943 (2021)
22. Guo, Y., Liu, J., Li, M., Tang, X., Chen, X., Zhao, B.: Vtg-llm: integrating timestamp knowledge into video llms for enhanced video temporal grounding. arXiv preprint arXiv:2405.13382 (2024)
23. Hao, J., Sun, H., Ren, P., Wang, J., Qi, Q., Liao, J.: Can shuffling video benefit temporal bias problem: a novel training framework for temporal grounding. In: European Conference on Computer Vision, pp. 130–147. Springer (2022)
24. He, Y., Yang, T., Zhang, Y., Shan, Y., Chen, Q.: Latent video diffusion models for high-fidelity video generation with arbitrary lengths. arXiv preprint arXiv:2211.13221 (2022)
25. Hendrycks, D., Basart, S., Mu, N., Kadavath, S., Wang, F., Dorundo, E., Desai, R., Zhu, T., Parajuli, S., Guo, M., et al.: The many faces of robustness: a critical analysis of out-of-distribution generalization. In: Proceedings of the IEEE/CVF International Conference on Computer Vision, pp. 8340–8349 (2021)

26. Ho, J., Salimans, T., Gritsenko, A., Chan, W., Norouzi, M., Fleet, D.J.: Video diffusion models. arXiv preprint arXiv:2204.03458 (2022)
27. Honnibal, M., Montani, I.: spacy 2: natural language understanding with bloom embeddings, convolutional neural networks and incremental parsing (2017). https://github.com/spacy-dev/Spacy2
28. Hu, Z., Ma, X., Liu, Z., Hovy, E., Xing, E.: Harnessing deep neural networks with logic rules. arXiv preprint arXiv:1603.06318 (2016)
29. Hu, R., Andreas, J., Rohrbach, M., Darrell, T., Saenko, K.: Learning to reason: end-to-end module networks for visual question answering. In: IEEE International Conference on Computer Vision, ICCV 2017, Venice, October 22–29, 2017, pp. 804–813 (2017)
30. Hu, R., Andreas, J., Darrell, T., Saenko, K.: Explainable neural computation via stack neural module networks. In: Proceedings of the European Conference on Computer Vision (ECCV), pp. 53–69 (2018)
31. Hu, E.J., Shen, Y., Wallis, P., Allen-Zhu, Z., Li, Y., Wang, S., Wang, L., Chen, W.: Lora: low-rank adaptation of large language models. arXiv preprint arXiv:2106.09685 (2021)
32. Hudson, D.A., Manning, C.D.: Compositional attention networks for machine reasoning. arXiv preprint arXiv:1803.03067 (2018)
33. Jia, M., Tang, L., Chen, B.C., Cardie, C., Belongie, S., Hariharan, B., Lim, S.N.: Visual prompt tuning. In: European Conference on Computer Vision, pp. 709–727. Springer (2022)
34. Johnson, J., Hariharan, B., van der Maaten, L., Hoffman, J., Fei-Fei, L., Zitnick, C.L., Girshick, R.B.: Inferring and executing programs for visual reasoning. In: IEEE International Conference on Computer Vision, ICCV 2017, Venice, October 22–29, 2017, pp. 3008–3017 (2017)
35. Jung, M., Jang, Y., Choi, S., Kim, J., Kim, J.H., Zhang, B.T.: Overcoming weak visual-textual alignment for video moment retrieval. arXiv preprint arXiv:2306.02728 (2023)
36. Khachatryan, L., Movsisyan, A., Tadevosyan, V., Henschel, R., Wang, Z., Navasardyan, S., Shi, H.: Text2video-zero: text-to-image diffusion models are zero-shot video generators. arXiv preprint arXiv:2303.13439 (2023)
37. Kim, D., Park, J., Lee, J., Park, S., Sohn, K.: Language-free training for zero-shot video grounding. arXiv preprint arXiv:2210.12977 (2022)
38. Krueger, D., Caballero, E., Jacobsen, J.H., Zhang, A., Binas, J., Zhang, D., Le Priol, R., Courville, A.: Out-of-distribution generalization via risk extrapolation (rex). In: International Conference on Machine Learning, pp. 5815–5826. PMLR (2021)
39. Lan, X., Yuan, Y., Wang, X., Chen, L., Wang, Z., Ma, L., Zhu, W.: A closer look at debiased temporal sentence grounding in videos: dataset, metric, and approach. ACM Trans. Multimedia Comput. Commun. Appl. **19**(6), 1–23 (2022)
40. Lan, X., Yuan, Y., Chen, H., Wang, X., Jie, Z., Ma, L., Wang, Z., Zhu, W.: Curriculum multi-negative augmentation for debiased video grounding. In: Proceedings of the AAAI Conference on Artificial Intelligence (2023)
41. Lester, B., Al-Rfou, R., Constant, N.: The power of scale for parameter-efficient prompt tuning. arXiv preprint arXiv:2104.08691 (2021)
42. Li, X.L., Liang, P.: Prefix-tuning: optimizing continuous prompts for generation. arXiv preprint arXiv:2101.00190 (2021)
43. Li, G., Wang, X., Zhu, W.: Perceptual visual reasoning with knowledge propagation. In: Proceedings of the 27th ACM International Conference on Multimedia, pp. 530–538 (2019)
44. Li, X., Dai, Y., Ge, Y., Liu, J., Shan, Y., DUAN, L.: Uncertainty modeling for out-of-distribution generalization. In: International Conference on Learning Representations (2021)
45. Li, J., Xie, J., Qian, L., Zhu, L., Tang, S., Wu, F., Yang, Y., Zhuang, Y., Wang, X.E.: Compositional temporal grounding with structured variational cross-graph correspondence learning. In: Proceedings of the IEEE/CVF Conference on Computer Vision and Pattern Recognition, pp. 3032–3041 (2022)
46. Li, J., Li, D., Savarese, S., Hoi, S.: Blip-2: bootstrapping language-image pre-training with frozen image encoders and large language models. In: International Conference on Machine Learning, pp. 19730–19742. PMLR (2023)

47. Li, J., Tang, S., Zhu, L., Zhang, W., Yang, Y., Chua, T.S., Wu, F.: Variational cross-graph reasoning and adaptive structured semantics learning for compositional temporal grounding. IEEE Trans. Pattern Anal. Mach. Intell. **45**(10), 12601–12617 (2023)
48. Lin, T.Y., Goyal, P., Girshick, R., He, K., Dollár, P.: Focal loss for dense object detection. In: Proceedings of the IEEE International Conference on Computer Vision, pp. 2980–2988 (2017)
49. Lindenbaum, O., Stanley, J., Wolf, G., Krishnaswamy, S.: Geometry based data generation. Adv. Neural Inf. Process. Syst. **31**, 1407–1418 (2018)
50. Liu, B., Yeung, S., Chou, E., Huang, D.A., Fei-Fei, L., Niebles, J.C.: Temporal modular networks for retrieving complex compositional activities in videos. In: Proceedings of the European Conference on Computer Vision (ECCV), pp. 552–568 (2018)
51. Liu, Z., Miao, Z., Zhan, X., Wang, J., Gong, B., Yu, S.X.: Large-scale long-tailed recognition in an open world. In: Proceedings of the IEEE/CVF Conference on Computer Vision and Pattern Recognition, pp. 2537–2546 (2019)
52. Liu, X., Ji, K., Fu, Y., Tam, W.L., Du, Z., Yang, Z., Tang, J.: P-tuning v2: prompt tuning can be comparable to fine-tuning universally across scales and tasks. arXiv preprint arXiv:2110.07602 (2021)
53. Liu, D., Qu, X., Hu, W.: Reducing the vision and language bias for temporal sentence grounding. In: Proceedings of the 30th ACM International Conference on Multimedia, pp. 4092–4101 (2022)
54. Liu, H., Li, C., Li, Y., Lee, Y.J.: Improved baselines with visual instruction tuning. In: Proceedings of the IEEE/CVF Conference on Computer Vision and Pattern Recognition (CVPR), pp. 26296–26306 (2024)
55. Lu, H., Liu, W., Zhang, B., Wang, B., Dong, K., Liu, B., Sun, J., Ren, T., Li, Z., Sun, Y., Deng, C., Xu, H., Xie, Z., Ruan, C.: Deepseek-vl: towards real-world vision-language understanding (2024). https://github.com/deepseek-ai/DeepSeek-VL
56. Luo, H., Ji, L., Zhong, M., Chen, Y., Lei, W., Duan, N., Li, T.: Clip4clip: an empirical study of clip for end to end video clip retrieval and captioning. Neurocomputing **508**, 293–304 (2022)
57. Luo, D., Huang, J., Gong, S., Jin, H., Liu, Y.: Towards generalisable video moment retrieval: visual-dynamic injection to image-text pre-training. In: Proceedings of the IEEE/CVF Conference on Computer Vision and Pattern Recognition, pp. 23045–23055 (2023)
58. Luo, R., Zhao, Z., Yang, M., Dong, J., Qiu, M., Lu, P., Wang, T., Wei, Z.: Valley: video assistant with large language model enhanced ability. arXiv preprint arXiv:2306.07207 (2023)
59. Luo, Z., Chen, D., Zhang, Y., Huang, Y., Wang, L., Shen, Y., Zhao, D., Zhou, J., Tan, T.: Videofusion: decomposed diffusion models for high-quality video generation. In: Proceedings of the IEEE/CVF Conference on Computer Vision and Pattern Recognition (2023)
60. Manhaeve, R., Dumancic, S., Kimmig, A., Demeester, T., De Raedt, L.: Deepproblog: neural probabilistic logic programming. Adv. Neural Inf. Process. Syst. **31**, 3753–3763 (2018)
61. Mao, J., Gan, C., Kohli, P., Tenenbaum, J.B., Wu, J.: The neuro-symbolic concept learner: interpreting scenes, words, and sentences from natural supervision. arXiv preprint arXiv:1904.12584 (2019)
62. Mascharka, D., Tran, P., Soklaski, R., Majumdar, A.: Transparency by design: Closing the gap between performance and interpretability in visual reasoning. In: Proceedings of the IEEE Conference on Computer Vision and Pattern Recognition, pp. 4942–4950 (2018)
63. Nam, J., Ahn, D., Kang, D., Ha, S.J., Choi, J.: Zero-shot natural language video localization. In: Proceedings of the IEEE/CVF International Conference on Computer Vision, pp. 1470–1479 (2021)
64. Nan, G., Qiao, R., Xiao, Y., Liu, J., Leng, S., Zhang, H., Lu, W.: Interventional video grounding with dual contrastive learning. In: Proceedings of the IEEE/CVF Conference on Computer Vision and Pattern Recognition, pp. 2765–2775 (2021)
65. Otani, M., Nakashima, Y., Rahtu, E., Heikkilä, J.: Uncovering hidden challenges in query-based video moment retrieval. In: 31st British Machine Vision Conference 2020, BMVC 2020, Virtual Event, September 7–10, 2020 (2020)

66. Paul, S., Mithun, N.C., Roy-Chowdhury, A.K.: Text-based temporal localization of novel events. In: Computer Vision–ECCV 2022: 17th European Conference, Tel Aviv, October 23–27, 2022, Proceedings, Part XIV, pp. 567–587. Springer (2022)
67. Peng, Z., Wang, W., Dong, L., Hao, Y., Huang, S., Ma, S., Wei, F.: Kosmos-2: grounding multimodal large language models to the world. ArXiv abs/2306 (2023)
68. Qu, M., Tang, J.: Probabilistic logic neural networks for reasoning. Adv. Neural Inf. Process. Syst. **32**, 7712–7722 (2019)
69. Qu, M., Chen, X., Liu, W., Li, A., Zhao, Y.: Chatvtg: video temporal grounding via chat with video dialogue large language models. In: Proceedings of the IEEE/CVF Conference on Computer Vision and Pattern Recognition, pp. 1847–1856 (2024)
70. Radford, A., Wu, J., Child, R., Luan, D., Amodei, D., Sutskever, I., et al.: Language models are unsupervised multitask learners. OpenAI Blog **1**(8), 9 (2019)
71. Radford, A., Kim, J.W., Hallacy, C., Ramesh, A., Goh, G., Agarwal, S., Sastry, G., Askell, A., Mishkin, P., Clark, J., et al.: Learning transferable visual models from natural language supervision. In: International Conference on Machine Learning, pp. 8748–8763. PMLR (2021)
72. Rombach, R., Blattmann, A., Lorenz, D., Esser, P., Ommer, B.: High-resolution image synthesis with latent diffusion models. In: Proceedings of the IEEE/CVF Conference on Computer Vision and Pattern Recognition, pp. 10684–10695 (2022)
73. Sikka, K., Huang, J., Silberfarb, A., Nayak, P., Rohrer, L., Sahu, P., Byrnes, J., Divakaran, A., Rohwer, R.: Zero-shot learning with knowledge enhanced visual semantic embeddings. arXiv preprint arXiv:2011.10889 (2020)
74. Soldan, M., Pardo, A., Alcázar, J.L., Caba, F., Zhao, C., Giancola, S., Ghanem, B.: Mad: a scalable dataset for language grounding in videos from movie audio descriptions. In: Proceedings of the IEEE/CVF Conference on Computer Vision and Pattern Recognition, pp. 5026–5035 (2022)
75. Tang, Y., Bi, J., Xu, S., Song, L., Liang, S., Wang, T., Zhang, D., An, J., Lin, J., Zhu, R., Vosoughi, A., Huang, C., Zhang, Z., Zheng, F., Zhang, J., Luo, P., Luo, J., Xu, C.: Video understanding with large language models: a survey. arXiv preprint arXiv:2312.17432 (2023)
76. Touvron, H., Lavril, T., Izacard, G., Martinet, X., Lachaux, M.A., Lacroix, T., Rozière, B., Goyal, N., Hambro, E., Azhar, F., et al.: Llama: open and efficient foundation language models. arXiv preprint arXiv:2302.13971 (2023)
77. Wang, X., Ye, Y., Gupta, A.: Zero-shot recognition via semantic embeddings and knowledge graphs. In: Proceedings of the IEEE Conference on Computer Vision and Pattern Recognition, pp. 6857–6866 (2018)
78. Wang, G., Wu, X., Liu, Z., Yan, J.: Prompt-based zero-shot video moment retrieval. In: Proceedings of the 30th ACM International Conference on Multimedia, pp. 413–421 (2022)
79. Wang, J., Lan, C., Liu, C., Ouyang, Y., Qin, T., Lu, W., Chen, Y., Zeng, W., Yu, P.: Generalizing to unseen domains: a survey on domain generalization. IEEE Trans. Knowl. Data Eng. **35**, 8052–8072 (2022)
80. Wang, J., Chen, D., Luo, C., Dai, X., Yuan, L., Wu, Z., Jiang, Y.G.: Chatvideo: a tracklet-centric multimodal and versatile video understanding system. arXiv preprint arXiv:2304.14407 (2023)
81. Xu, G., Duong, T.D., Li, Q., Liu, S., Wang, X.: Causality learning: a new perspective for interpretable machine learning. arXiv preprint arXiv:2006.16789 (2020)
82. Yang, X., Feng, F., Ji, W., Wang, M., Chua, T.S.: Deconfounded video moment retrieval with causal intervention. In: Proceedings of the 44th International ACM SIGIR Conference on Research and Development in Information Retrieval, SIGIR '21, pp. 1–10 (2021)
83. Yang, L., Kong, Q., Yang, H.K., Kehl, W., Sato, Y., Kobori, N.: Deco: decomposition and reconstruction for compositional temporal grounding via coarse-to-fine contrastive ranking. In: Proceedings of the IEEE/CVF Conference on Computer Vision and Pattern Recognition, pp. 23130–23140 (2023)
84. Ye, H., Xie, C., Cai, T., Li, R., Li, Z., Wang, L.: Towards a theoretical framework of out-of-distribution generalization. Adv. Neural Inf. Process. Syst. **34**, 23519–23531 (2021)

85. Yenduri, G., Ramalingam, M., Selvi, G.C., Supriya, Y., Srivastava, G., Maddikunta, P.K.R., Raj, G.D., Jhaveri, R.H., Prabadevi, B., Wang, W., et al.: Gpt (generative pre-trained transformer)–a comprehensive review on enabling technologies, potential applications, emerging challenges, and future directions. IEEE Access (2024)
86. Yu, D., Yang, B., Wei, Q., Li, A., Pan, S.: A probabilistic graphical model based on neural-symbolic reasoning for visual relationship detection. In: Proceedings of the IEEE/CVF Conference on Computer Vision and Pattern Recognition, pp. 10609–10618 (2022)
87. Yu, D., Yang, B., Liu, D., Wang, H., Pan, S.: A survey on neural-symbolic learning systems. Neural Netw. **166**, 105–126 (2023)
88. Yu, E., Zhao, L., Wei, Y., Yang, J., Wu, D., Kong, L., Wei, H., Wang, T., Ge, Z., Zhang, X., et al.: Merlin: empowering multimodal llms with foresight minds. arXiv preprint arXiv:2312.00589 (2023)
89. Yuan, Y., Lan, X., Wang, X., Chen, L., Wang, Z., Zhu, W.: A closer look at temporal sentence grounding in videos: dataset and metric. In: Proceedings of the 2nd International Workshop on Human-Centric Multimedia Analysis, pp. 13–21 (2021)
90. Zhang, H., Sun, A., Jing, W., Zhou, J.T.: Towards debiasing temporal sentence grounding in video. arXiv preprint arXiv:2111.04321 (2021)
91. Zhang, X., Cui, P., Xu, R., Zhou, L., He, Y., Shen, Z.: Deep stable learning for out-of-distribution generalization. In: Proceedings of the IEEE/CVF Conference on Computer Vision and Pattern Recognition, pp. 5372–5382 (2021)
92. Zhao, Y., Misra, I., Krähenbühl, P., Girdhar, R.: Learning video representations from large language models. In: Proceedings of the IEEE/CVF Conference on Computer Vision and Pattern Recognition, pp. 6586–6597 (2023)
93. Zhou, Z.H.: Abductive learning: towards bridging machine learning and logical reasoning. Sci. China Inf. Sci. **62**(7), 76101 (2019)
94. Zhou, D., Wang, W., Yan, H., Lv, W., Zhu, Y., Feng, J.: Magicvideo: efficient video generation with latent diffusion models. arXiv preprint arXiv:2211.11018 (2022)
95. Zhu, D., Chen, J., Shen, X., Li, X., Elhoseiny, M.: Minigpt-4: Enhancing vision-language understanding with advanced large language models. arXiv preprint arXiv:2304.10592 (2023)

Index

X. Wang et al., *Video Grounding and Its Generalization*,
https://doi.org/10.1007/978-3-031-94837-4

MIX
Papier aus verantwortungsvollen Quellen
Paper from responsible sources
FSC® C105338

If you have any concerns about our products,
you can contact us on
ProductSafety@springernature.com

In case Publisher is established outside the EU,
the EU authorized representative is:
Springer Nature Customer Service Center GmbH
Europaplatz 3, 69115 Heidelberg, Germany

Printed by Libri Plureos GmbH
in Hamburg, Germany